second edition
Chemical Genomics
and Proteomics

Edited by
Ferenc Darvas • András Guttman • György Dormán

second edition
Chemical Genomics
and Proteomics

CRC Press
Taylor & Francis Group
Boca Raton   London   New York

CRC Press is an imprint of the
Taylor & Francis Group, an **informa** business

CRC Press
Taylor & Francis Group
6000 Broken Sound Parkway NW, Suite 300
Boca Raton, FL 33487-2742

First issued in paperback 2016

ISBN 13: 978-1-138-19847-0 (pbk)
ISBN 13: 978-1-4398-3052-9 (hbk)

Library of Congress Cataloging-in-Publication Data

Chemical genomics and proteomics / editors, Ferenc Darvas, András Guttman, György Dormán. -- 2nd ed.
    p. cm.
    Rev. ed. of: Chemical genomics. New York : Marcel Dekker, c2004.
    Includes bibliographical references and index.
    ISBN 978-1-4398-3052-9 (hardcover : alk. paper)
    1. Chemogenomics. 2. Proteomics. 3. Pharmacogenetics. 4. Biochemical genetics. 5. DNA microarrays. 6. Combinatorial chemistry. I. Darvas, F. II. Guttman, András. III. Dormán, G. (György) IV. Chemical genomics.

QH431.C45196 2013
615.1'9--dc23   2012028129

Visit the Taylor & Francis Web site at
http://www.taylorandfrancis.com

and the CRC Press Web site at
http://www.crcpress.com

Contents

Preface

In the late 1990s, chemical genetics emerged to study the interaction or perturbation of cells with small molecules in order to mimic the effect of mutations investigated by functional genomics. While classical genetics shows particular advantages over chemical genetics regarding the high specificity that can be achieved with gene knockouts, chemical genetics could only meet this challenge if it could dissect the function of organisms and cells by having a specific small molecule partner for every gene product. While chemical genetics studies the changes of individual genes or one particular phenotype induced by small molecules, chemical genomics investigates the alteration of the whole genome *en masse* through the perturbation of the complex network of signaling pathways by small molecules; thus, chemical genomics is an extension of chemical genetics to a genome-wide scale. This basic concept found several innovative applications over the years from cell biology, to drug discovery, and more recently in diagnostics and clinical practice.

The pioneering first edition of this book, *Chemical Genomics*, was published by Marcel Dekker, Inc., more than 7 years ago. Since then, the area of chemical genomics has rapidly expanded and diversified; numerous novel methods and subdisciplines such as (chemical) glycomics and lipidomics have emerged as well, showing interdisciplinary features.

Chapter 1, while giving a short account of the general strategies and methodologies of chemical genomics, mainly focuses on the emerging technologies and recent applications in the areas of combination chemical genetics, toxicogenomics, drug chemical genomics/proteomics, and orthogonal chemical genetics, which utilizes small molecules and/or compound libraries.

The integration of cell-based assays with activity/affinity-based approaches allows us to identify the proteins targeted by the small molecules, the altered expression levels, and ultimately reveals the association with signaling pathways or disease states. Chemical genomics is closely related to the novel postgenomic discovery strategy, where target identification/validation is the key element. All these efforts help to understand the cell functions as well as the pathomechanisms of diseases, which could open new opportunities for therapeutic interventions.

Chapter 2 discusses the development and application of novel analytical methods that are used in lipidomics (including steroidomics), which is one of the metabolomics approaches. Metabolomics, which is the systematic identification of the composition of low molecular weight materials within the cells/body upon small-molecule treatment or induced by disease, is closely related to chemical genomics since the direct binding interaction with proteins within the cells could also lead to changing the concentration of the small molecule metabolites. The primary objectives of lipidomics are to determine the quantitative lipid profile of a cell type, tissue, or biological fluid, and mass spectrometry is the dominant analytical technique together with its variants (shotgun lipidomics, gas or liquid chromatography, mass spectrometry, tandem mass spectrometry).

Since the cellular response (gene expression alterations or phenotypic changes) to small molecules can be significantly different in normal and in pathological states, differential omics technologies allow for the identification of gene or protein biomarkers.

In Chapter 3, biomarker discovery applications of microarray technologies using DNA, RNA, and protein and glycan arrays, with the advantage of requiring very small amounts of samples with excellent detection limits, are discussed. Microarray technology-based approaches have been applied in serum tumor marker profiling as well as in drug discovery and medicinal chemistry when the effects of small molecules in DNA–protein or protein–protein interaction are studied.

Chapter 4 describes further applications of biomolecular biomarkers often used in the diagnosis of diseases, at various stages in the small molecule drug R&D process and during the therapeutic use of marketed medicines. Throughout the process, the use of biomarkers is intended to secure the highest level of efficiency and safety possible for a given drug, that is, during the drug discovery phase to select the best chemical entity that hits the target or during the drug development phase, to define the target patient group which will respond to treatment. The major types of biomarkers are discussed in detail, for example, prognostic biomarkers, disease-specific biomarkers, response (surrogate) biomarkers, and toxicity biomarkers.

Since chemical genomics represents a holistic approach, in principle, the small molecules can interact with the whole proteome either through direct binding interaction or indirectly as a downstream effect through the signaling network. Furthermore, the cell response is the overall consequence of the interactions initiated by the small molecules, thus, chemical genomics is closely associated with systems biology and network theory.

The revolutionary development of chemical and functional omics approaches (genomics, proteomics, lipidomics, glycomics, etc.) has contributed to the emergence of database management, high-throughput molecular technologies, and multivariant bioinformatic analysis technologies leading to an entire change in the paradigm of experimental biology and medicine. This new paradigm in biomedical sciences includes a network-oriented view and advanced statistical and informatics data management, which is based on network theories and mathematical modeling and opens the way toward personalized medicine. In Chapter 5, the principles of contemporary systems biology and genomics in experimental medicine are discussed.

Finally, the chemical genomics/proteomics methods are complemented by a much broader discipline called *chemogenomics*, which is in close relationship with system biology and investigates the matching of the chemical and biological space integrating experimental and *in silico* approaches. Chemogenomics is a recent global drug discovery paradigm integrating the knowledge of gene sequences, protein families, and protein–small molecule interaction based on experimental and 2D/3D molecular modeling.

Chapter 6 discusses the recent recognition that the significant attrition rate in clinical trials during the past decades has been due to the unexpected binding of drug candidates to additional targets, either off-targets or antitargets resulting in dubious pharmacological activities, side effects, and sometimes adverse drug reactions. In this chapter, various *in silico* chemogenomics approaches are discussed, which utilize annotated ligand-target as well as protein–ligand interaction databases, and could predict or reveal promiscuous binding and polypharmacology, the knowledge of which would help medicinal chemists in designing more potent clinical candidates with fewer side effects.

The book is supplemented with a glossary and a basic index.

In summary, the selected topics of *Chemical Genomics and Proteomics*, 2nd Edition, reflects the major directions of the field witnessed in the last couple of years together with the recently emerged interdisciplinary aspects of chemical genomics demonstrated by various applications. The target audience of this book has also been extended; not only to practitioners in drug discovery, medicinal chemistry, and molecular/cell biology, but also to physicians in clinical diagnostics as well as experts in the whole drug development pipeline, who could find valuable information in this second edition. This edition is equally useful for scientists working in laboratories and for students at the graduate and undergraduate levels, and it serves as a valuable textbook for university professors offering courses on medicinal chemistry, pharmacy, and medicine in the postgenomic era.

Ferenc Darvas
András Guttman
György Dormán

Contributors

Peter Antal
Abiomics Ltd.
Budapest, Hungary

Edit Buzás
Department of Genetics, Cell, and
  Immunobiology
Semmelweis University
and
Research Group of Inflammation
  Biology and Immunogenomics
Hungarian Academy of Sciences
  and Semmelweis University
Budapest, Hungary

Ferenc Darvas
ThalesNano Inc.
Budapest, Hungary
and
Florida International University
Miami, Florida

Anna Debreceni
Biosystems International Kft.
Debrecen, Hungary

György Dormán
ThalesNano Inc.
Budapest, Hungary
and
University of Szeged
Szeged, Hungary

Katalin Éder
Department of Genetics, Cell, and
  Immunobiology
Semmelweis University
and
Research Group of Inflammation
  Biology and Immunogenomics
Hungarian Academy of Sciences
  and Semmelweis University
Budapest, Hungary

András Falus
Department of Genetics, Cell, and
  Immunobiology
Semmelweis University
and
Research Group of Inflammation
  Biology and Immunogenomics
Hungarian Academy of Sciences
  and Semmelweis University
Budapest, Hungary

William J. Griffiths
Institute of Mass Spectrometry
School of Medicine
Swansea University
Swansea, United Kingdom

András Guttman
Horváth Laboratory of
Bioseparation Sciences
University of Debrecen
Debrecen, Hungary
and

The Barnett Institute
Northeastern University
Boston, Massachusetts

István Kurucz
Biosystems International Kft.
Debrecen, Hungary

Anna Meljon
Institute of Mass Spectrometry
School of Medicine
Swansea University
Swansea, United Kingdom

Viktor Molnar
Department of Genetics, Cell, and
  Immunobiology
Semmelweis University
and
Research Group of Inflammation
  Biology and Immunogenomics
Hungarian Academy of Sciences
  and Semmelweis University
Budapest, Hungary

Michael Ogundare
Institute of Mass Spectrometry
School of Medicine
Swansea University
Swansea, United Kingdom

Éva Pállinger
Department of Genetics, Cell, and
  Immunobiology
Semmelweis University
and
Research Group of Inflammation
  Biology and Immunogenomics
Hungarian Academy of Sciences
  and Semmelweis University
Budapest, Hungary

Didier Rognan
Structural Chemogenomics
Université de Strasbourg
Illkirch, France

Sándor Spisák
Horváth Laboratory of
  Bioseparation Sciences
University of Debrecen
Debrecen, Hungary
and
The Barnett Institute
Northeastern University
Boston, Massachusetts

Tamás Szabó
Department of Genetics, Cell, and
  Immunobiology
Semmelweis University
and
Research Group of Inflammation
  Biology and Immunogenomics
Hungarian Academy of Sciences
  and Semmelweis University
Budapest, Hungary

Csaba Szalai
Abiomics Ltd.
Budapest, Hungary

László Takács
Biosystems International Kft.
and
Department of Human Genetics
Medical and Health Science Center
University of Debrecen
Debrecen, Hungary

Gergely Tölgyesi
Department of Genetics, Cell, and
  Immunobiology
Semmelweis University
and

Research Group of Inflammation
  Biology and Immunogenomics
Hungarian Academy of Sciences
  and Semmelweis University
Budapest, Hungary

Yuqin Wang
Institute of Mass Spectrometry
School of Medicine
Swansea University
Swansea, United Kingdom

chapter one

Utilizing small molecules in chemical genomics
Toward high-throughput (HT) approaches

György Dormán and Ferenc Darvas

Contents

1.1   Introduction

The idea of using small molecules in complex organisms and cell-based assays to discover biologically active compounds is one of the earliest concepts in medicinal chemistry and classical pharmacology. In the postgenomic era after the completion of the sequencing stage of the Human Genome Project, high-throughput (HT) cell-based biological screening systems as well as DNA microarrays were developed which allow detecting any interactions with the whole genome or proteome through phenotypic or gene expression changes. Furthermore, the physiological effects of the small molecules can also be investigated through gene or protein expression.

Subsequently, the binding partners of the small molecule (molecular targets) can be identified which helps to understand the cell functions as well as pathomechanisms of the diseases, which could open up new opportunities for therapeutic interventions.

While chemical genetics studies the changes of individual genes or one particular phenotype induced by small molecules, chemical genomics investigates the alteration of the whole genome *en masse* through the perturbation of the complex network of signaling pathways by small molecules, thus, chemical genomics is an extension of chemical genetics to a genome-wide scale (MacBeath, 2001). (*Omics* means to study something on a large scale systematically, rather than one-by-one and on a case-by-case basis). The detected genome-wide phenotypic and/or gene expression changes could reveal the functions of genes/proteins or gene/protein families.

Chemical genomics represents a holistic approach since in principle the small molecules could interact with the whole proteome either through direct binding interaction or indirectly as a downstream effect through the signaling network. Since the cell response is the overall consequence of the interactions initiated by the small molecules, chemical genomics is closely associated with systems biology and network theory (Yildirim et al., 2007).

These novel disciplines examine the behavior of complex (biological) systems upon perturbation with genetic, environmental, and chemical determinants such as small molecules (Butcher, Berg, and Kunkel, 2004).

While the small molecules are acting upon the binding to one or several different proteins at different stages of the signaling network, this binding initiates the signal transduction through releasing effector molecules (second messengers), activating enzymes (e.g., phosphorylation), and protein–protein interactions (interactome) leading to the cellular response (gene expression alterations or phenotypic changes), which can be significantly different in normal and in pathological states.

1.2   Small molecules in chemical genomics/proteomics

Treating the cells with small molecules can be studied at two levels (Figure 1.1) (Dormán and Darvas, 2004). At the cellular level, small molecules could interact with the signaling network/pathways through numerous potential protein targets within the cell from the membrane to the nucleus and the resulting consequences (total cell response) can be measured by gene expression, functional, phenotypic, or viability assays at the cellular level (Croston, 2002). Thus, the investigations are not restricted to any single cell-surface receptor or recombinant protein target participating in the cell-to-cell or intracellular communication.

In the pathological state, as a consequence of mutation or altered gene expression, the signaling networks/pathways are disrupted leading to

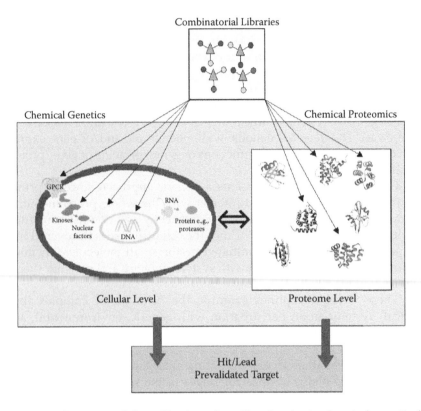

Figure 1.1 Overview of the utilization of small molecules in chemical genetics/ proteomics. **(See color insert.)**

typical phenotypic changes or gene expression patterns characteristic to the disease. Upon treatment with small molecules, monitoring and comparing the phenotypic changes or gene expression patterns in healthy and diseased states allow accurate diagnosis and evaluation of the progress of the disease (Stockwell, 2000) at the proteome level.

In order to identify the functions of the genes and gene products associated with the signaling network and rationalize the cellular responses, the proteome comprised by the cell should be probed with small molecules (Figure 1.1).

In fact, this is the genome-wide application of the target-based drug discovery and the extension of the principle of chemical genomics to proteomics.

The differences in gene expression within the cell at the proteome level can be measured and compared through activity/affinity-based interactions with well-designed small molecules.

Furthermore, the activity/affinity-based methods are also suitable for identifying or annotating the functions of the gene product proteins as well as their drug ability (Barglow and Cravatt, 2007).

In general, this approach is often called *chemical proteomics*, by definition a parallel investigation of the interactions between small molecule libraries and the proteome (proteins that constitute the cell in native or in isolated form in separate experience) at the proteome level.

Considering the drug discovery pipeline, this two-level chemical genomic/proteomic screening in an optimal case would result in a lead compound as well as prevalidated protein targets.

The chemical genomics/proteomics methods are complemented with a much broader discipline, called *chemogenomics*, which builds a relationship with the system biology and investigates the matching of the chemical and biological space integrating experimental and *in silico* approaches.

In other words, this is a new global drug discovery paradigm integrating the knowledge of gene sequences, protein families, and protein–small molecule interaction based on experimental and 2D/3D molecular modeling.

1.2.1 The advantages of chemical genetics/genomics

While the classical genetics experiments are difficult to carry out, particularly in mammalian organisms, chemical genetics is applicable in any cellular systems as well as in animal models. In the classical knockout experiments, the protein concerned is completely eliminated from the organism.

Since normally any protein has many different functions, by applying a highly selective inhibitor or modulator in a chemical genetics experiment, a particular function can be *dissected*; in other words investigated separately without affecting the others.

Furthermore, treating the cells with small molecules allows gradual changes in the cellular function due to concentration dependency; thus, a sublethal dose can also be applied while in genetics knocking out a gene has lethal consequences.

In summary, chemical genetics facilitates a real-time control in cellular studies and only diffusion could be a limitation.

On the other hand, classical genetics shows a particular advantage over chemical genetics regarding the high specificity that can be achieved with gene knockouts. Chemical genetics could only meet this challenge if it sets a very ambitious goal "to dissect the function of organism and cells by having a small molecule partner for every gene product" (Schreiber, 1998), which allows investigating one particular biological pathway without affecting others, while binding to its protein binding partner. This is the primary mission of chemical genetics.

The more selective is the binding to a particular target protein within the intracellular signaling network, the more efficiently the small molecule mimics the effect of a mutation in classical genetics.

To achieve this, we take advantage of the increased specificity of the natural products that were evolved through permanent interactions with target proteins in living organisms. Alternatively, the increased structural diversity of the synthetic small molecule libraries can be exploited.

1.2.2   Small molecules and synthetic approaches

1.2.2.1   Natural products and natural product-like compounds: Biology- and diversity-oriented synthesis

As stated earlier, a cell-based chemical genomics screening approach requires highly specific and tightly binding small molecules for each expressed protein participating in cell cycle pathways and regulation (Khersonsky and Chang, 2004). Such molecules can be used as probes for functional genomics, pharmacology, and finally as drug prototypes. Novel chemistry would be required, directed at the creation of highly diverse, more complex, small synthetic molecules, such as natural products. Furthermore, natural product collection is presumed to contain a much higher proportion of compounds that are likely to exhibit bioactivity, selectively modifying cellular functions, since these compounds have evolved for a specific reason in the living organism (from marine sponges, and plant and soil sources) for example, and defense against other organisms. They are frequently toxic and could be DNA cleaving agents. Comparing the structural features of the abundance of natural molecules, it is apparent that they provide a much greater structural variability and diversity than traditional synthetic *drug-like* molecules. Diversity-orientated chemistry yields molecules similar to those found in nature. Examples of efficient synthesis strategies toward *natural-like* chemical scaffolds have been established (Reayi and Arya, 2005). The combinatorial approaches toward natural product analogs were recently reviewed by Joseph and Arya (2004).

The synthesis and application of natural-like libraries were also justified by the difficulty in biological screening of natural extracts, and isolation and identification of the active ingredients.

On the other hand, the natural products or natural product fragments, particularly in optically active form, could serve as suitable starting points for library design.

The molecular scaffolds (ring systems) validated by nature or biology can be arranged into hierarchical clusters: Structural Classification of Natural Products (SCONP) (Figure 1.2), which could accelerate the design

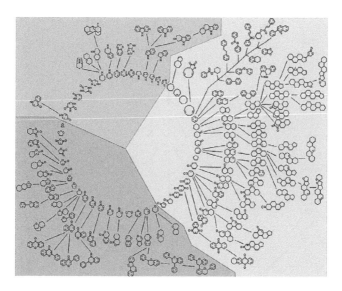

Figure 1.2 Structural Classification of Natural Products (SCONP): The molecular scaffolds (ring systems) derived from natural products arranged into hierarchical clusters. (From Bon, R. S. and Waldmann, H., 2010, Bioactivity-Guided Navigation of Chemical Space, *Acc. Chem. Res.*, 43(8):1103–14.)

of highly diverse natural product-like compound libraries (Schuffenhauer et al., 2007).

The library design and synthesis concept, which is based on the structural features of natural product scaffolds or reflects structural similarities to such scaffolds, is referred to as *biology-oriented synthesis* (BIOS) (Umarye et al., 2007). This approach expands the chemical space around the scaffolds of natural origin and allows a bioactivity-guided navigation (Bon and Waldmann, 2010). Figure 1.3 demonstrates this concept showing the coverage of the chemical space around the molecular scaffold of the natural product, yohimbine.

Another relevant library-building concept that also emerged from the diversity of natural products assumes that such molecular scaffolds primarily hold the biological activity, and the side chains and functional groups only modify or fine-tune the interaction. Therefore, a new term was introduced by the Schreiber group, *skeletal diversity* (Kwon, Park, and Schreiber, 2002). Instead of increasing the overall structural diversity or dissimilarity, skeletal diversity is required in libraries involved in chemical genetics which can only be achieved by *diversity-oriented synthesis* (DOS) (Arya et al., 2005; Tan, 2005).

The Diels–Alder reaction is particularly useful in DOS. Figure 1.4 shows two applications where natural product-like libraries with high skeletal diversity were obtained in two consecutive steps. The approach

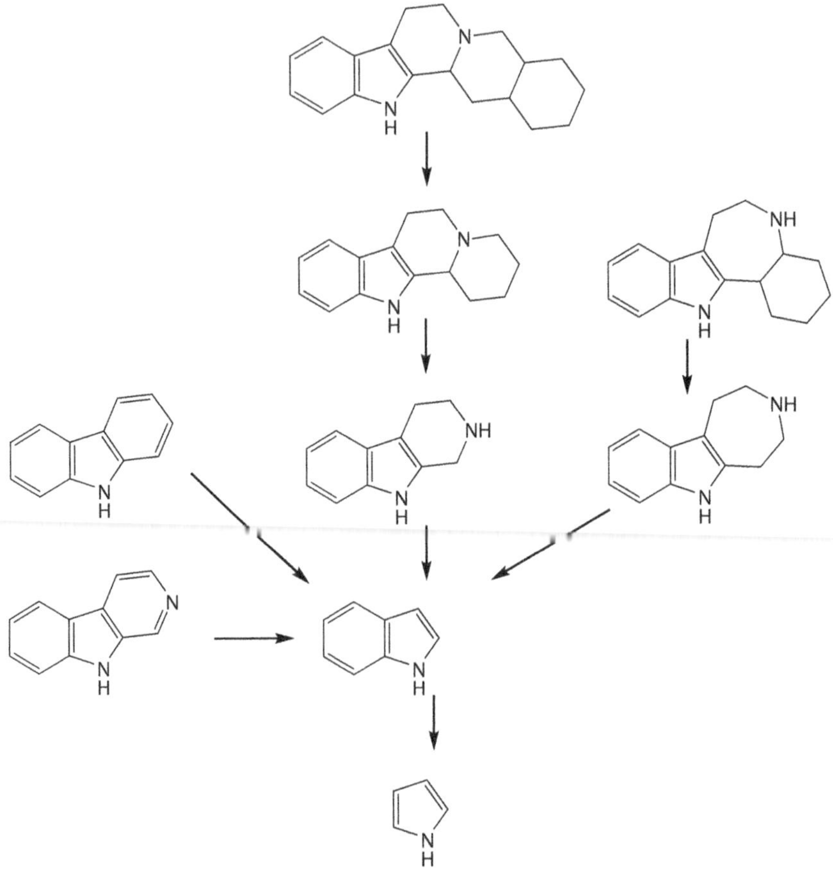

Figure 1.3 Exploring the chemical space around the molecular scaffold of yohimbine.

was used to prepare 29,400 polycyclic compounds around 10 distinct molecular skeletons.

DOS strategy is a significantly different synthetic concept from the traditional *target-oriented synthesis* and the difference is demonstrated in Figure 1.5. In target-oriented synthesis, one complex molecule is targeted and according to the retrosynthetic analysis, many different routes would lead to simple precursors. On the contrary, in DOS, the real target is the increased skeletal diversity (to cover the chemical space as much as possible) and not a single compound, which can be achieved with a minimum number of steps and with high economy in diversity building. DOS starts from simple multifunctional building blocks and linear, convergent, or branched synthetic routes yield compound libraries with diverse cores/skeletons applying novel and robust synthesis techniques.

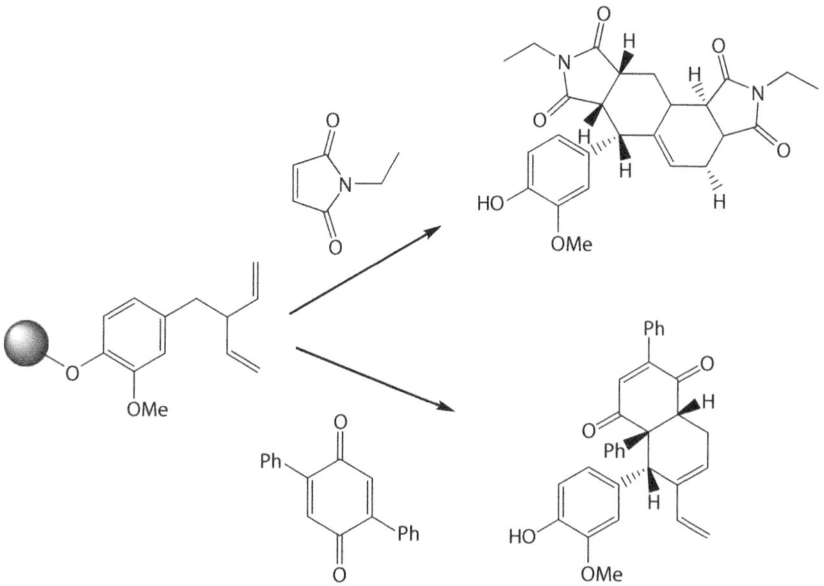

Figure 1.4 Examples of diversity-oriented synthesis (DOS) utilizing Diels–Alder reaction.

DOS aims to provide structurally complex and diverse small molecules efficiently for chemical genetic screens; thus, in a pioneering study, Schreiber and coworkers synthesized a polycyclic library that mimics the complexity of the biologically active molecules. Chemical genetic assays identified compounds that activate the TGF-beta responsive reporter gene (Tan et al., 1999) (Figure 1.6).

Recent reviews discuss the latest developments in DOS, which provide various examples leading to natural-like compound libraries (Cordier et al., 2008; Marcaurelle and Johannes, 2008). The *split-mix* solid-phase synthesis technology developed by Furka (Furka et al., 1991) is particularly useful for library generation using the DOS strategy leading to over a million members of libraries as mixtures.

1.2.2.2 *Structurally and mechanistically diverse libraries used for chemical genetics*

The total chemical space is estimated as 10^{180} distinct compounds (Mwt < 500 Da), although only 10^{60} can be considered as drug-like and this set is composed of 10^{40} different cores or scaffolds (Dean, 2007). From commercial sources, approximately 25 million compounds (ZINC database) (Irwin and Shoichet, 2005) are available although the diversity, particularly the skeletal diversity, is far behind what DOS strategy offers. On the

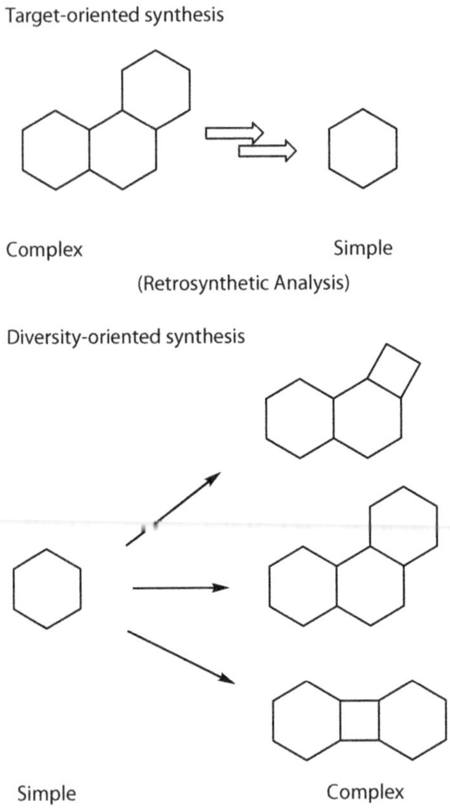

Target-oriented synthesis

Complex   Simple

(Retrosynthetic Analysis)

Diversity-oriented synthesis

Simple   Complex

Figure 1.5 Comparison of *target-oriented synthesis* and *diversity-oriented synthesis*.

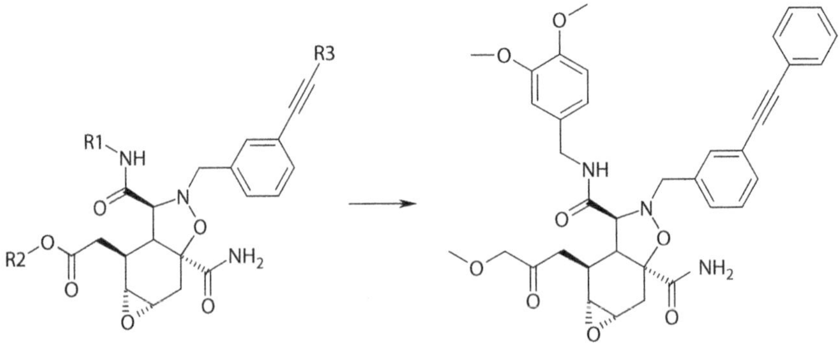

Figure 1.6 A *natural-like* polycyclic library for chemical genetics studies.

other hand, small molecules involved in cell-based chemical genetics screening should possess appropriate cell permeability, metabolic stability, and low general cytotoxicity. Several *in silico* methods are applied for filtering out the unsuitable library members for cell-based screening (Darvas et al., 2002).

In recent years, another approach was developed which focuses on biological or activity-based diversity in addition to chemical diversity in interrogating cells with small molecules in phenotypic or gene expression screens. Such so-called annotated libraries (Root et al., 2003) have high structural diversity and contain activity data from various cell-based or target-based biological screenings (from weak to strong activities) representing different pathways and mechanisms. Based on the results of the chemical genomics screens performed, annotated libraries could lead to identifying novel relationships between various intracellular pathways and to allowing the setting-up of novel hypotheses.

A similar concept uses chemogenomic reference libraries, which results in chemogenomics/toxicogenomics fingerprints for comparative studies.

For example DrugMatrix™, a postgenomic reference library array (Browne et al., 2002; Ganter et al., 2006), contains 1000 FDA approved drugs, 400 non-U.S. approved drugs, 400 standard biochemicals, 100 standard toxicants, and 100 withdrawn drugs with structural and mechanistic diversity. The library is employed in a chemical genomics reference system for chemogenomic profiling (expression array) at different stages of the discovery process (lead optimization, preclinical assessment, etc.) for expression profiling or phenotyping changes. The reference system generates a chemogenomic signature as well as markers characterizing a diseased state together with a detailed map of functional responses.

1.3   Global investigation of small molecule interactions with proteins that constitute the cell (proteome)

The interaction of small molecules with cells should also be investigated at the protein (proteome) level since the concentration (abundance), transcriptional alteration, posttranslational modification, formation of complexes with other proteins or cellular components, modulation of the activity with small molecule effectors, and compartmentation cannot be revealed at the genome level.

Determining the exact distribution, abundance, and expression level of the proteins within different cells is very important in identifying the role of a particular protein and its interactions within the cellular signaling pathways both in healthy and diseased states.

The human genome can be investigated at the proteome level through direct small molecule–gene product (protein) interactions using activity/affinity-based approaches. Such small molecule–protein interaction analysis serves several objectives and can be performed directly with cells, cell extracts, or alternatively applying isolated proteins in a parallel manner.

1.3.1 Druggability profiling of proteins using rapid affinity-based approaches

Investigation of the direct binding interactions between a large number of small molecules and the cca. 30,000 gene products enables a first filtering of the proteins as well as the small molecules. By definition, a protein target is druggable (chemically tractable) if it has an ability to bind a small molecule with an appropriate binding affinity. The small molecules that interact with at least one protein known to be involved in intracellular signaling could be applied in cell-based assays and this is likely to modulate the signaling network leading to cell response. On the other hand, in general, the binding proteins cannot be considered automatically as potential drug targets or targets for investigating cellular processes by chemical genomics until their role in the pathogenesis of a disease or in intracellular signaling is not confirmed or validated (Pal, 2000). It is very common for small molecules to bind to protein targets in a nonspecific manner since proteins normally offer numerous sites for binding. Even among the high-affinity interactions, there are several nonfunctional, secondary binding interactions.

As a result, a two-level definition of druggability is proposed:

1. At a primary level, a target is druggable if it has a specific binding partner.
2. At a secondary level, the target is druggable if its function can be modulated by small drug-like molecules. This level of druggability can be investigated in cell-based chemical genomics screens.

The major tool for druggability profiling is a rapid, uniform, high-throughput binding assay with a large, diverse compound library, often with small molecule microarrays.

For proteome mining/druggability or rapid affinity profiling, large diverse combinatorial libraries are employed either in a mixture or as individual compounds. These screens identify a certain target—a ligand affinity fingerprint that reveals the matching chemical and biological space (the objectives of chemogenomics) and allows an easy identification of a structure-activity relationship (SAR). Data mining is particularly easy if the compounds are immobilized on a miniaturized chip, and the fingerprint can be visualized easily either by fluorescence or surface plasmon resonance intensities in the different spots. The diverse libraries preferably contain various small molecules (fragments, drug motifs, or promiscuous structural elements) in addition to various physicochemical environments forming neutral or anionic species.

1.3.2 Fragment-based libraries for rapid affinity profiling

The composition of rapid affinity profiling libraries has been reported (Kessmann and Ottleben, 2001), based on information-rich building blocks (templates), and it is associated with fragment-based iterative screening approaches. Decreased size and complexity of the molecules (fragments) enables the structure's relative contribution to binding interactions to be identified, even if these fragments show low activity. The fragmental library design approach results in lead-like structures that are typically smaller and less hydrophobic than the drug-like compounds, as defined by Lipinski's Rule of 5 (Lipinski et al., 1997). The active, lead-like compounds could be optimized rapidly by adding chemical groups at open functionalities.

In 1999, Oprea proposed the concept of *lead-like combinatorial libraries* containing compounds with properties suitable for further optimization (Mwt: 100–350 and clog P 1–3) (Oprea and Gottfries, 1999). In fragment-based drug discovery, *in vitro* screening followed by fragment assembly could lead to compounds that occupy multiple binding domains leading to increased selectivity (Erlanson, 2006).

Rapid affinity profiling of the proteome could result in identifying individual compounds with high specificity against certain targets that can be involved in signaling pathways. Such compounds are potential candidates as reversible chemical knockouts to perturb the signaling network leading to a particular cellular response (gene expression or phenotypic alterations).

1.3.3 Functional profiling of the proteome

Proteome profiling is based on global activity screening (activity-based probes) (Hagenstein and Sewald, 2006). For the functional annotation of broad-spectrum enzymes, mechanism-based irreversible inhibitors (Evans and Cravatt, 2006) were designed for isolating and identifying members of gene (enzyme) families having the same function (e.g., catalytic activity). Such mechanism-based inhibitors (including protein-reactive natural products) (Drahl, Cravatt, and Sorensen, 2005) termed *activity-based probes* (Cravatt, Wright, and Kozarich, 2008) specifically target the active site of the enzyme resulting in a stable covalent linkage between the small molecules and the enzyme for separating a particular class of proteins

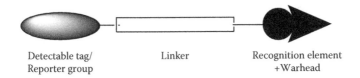

Detectable tag/   Linker   Recognition element
Reporter group   +Warhead

Figure 1.7 The major structural elements of ABPs.

from the entire proteome (*focusing*), thus reducing its complexity. Such reduction also improves the detection sensitivity of the expression profiling within the enzyme family. This covalent-bond forming method also allows the isolation of individual proteins for sequencing and identifying under denaturing conditions.

Activity-based probes (ABPs) contain: (1) a warhead—a class-selective electrophile that reacts primarily with a key amino acid in the active site of the proteins of a particular class and covalently attaches to it, plus a recognition element that directs the warhead more specifically to the active site; (2) a reporter tag that allows detection and isolation; and (3) a linker, which connects the elements with a proper distance and flexibility (Figure 1.7). ABPs have the potential to:

- Identify the members of a given enzyme family or activity class.
- Determine the catalytic activity levels of individual family members (abundance expression level).
- Localize active enzymes within cells.
- Screen small molecule libraries in crude protein extracts for inhibitors for comparative chemical proteomics studies (Paulick and Bogyo, 2008) (identifying differences in the expression profile of various enzymes in diseased state and healthy state).

These activity-based probes were successfully applied to identify major enzyme families like kinases, (e.g., serine/threonine and tyrosine kinases), phosphatases, and proteases (e.g., serine, aspartic, cysteine proteases etc.). In order to study enzymes where ABPs do not exist, a relatively classic approach provides opportunities, namely photocovalent cross-linking or photoaffinity labeling (Hatanaka and Sadakane, 2002).

1.3.3.1 *Posttranslational modification profiling*

Specific biochemical methods are used to identify the abundance of targets with specific posttranslational modifications (such as phosphorylated proteins) (Zheng and Chan, 2002). Monitoring protein phosphorylations is particularly important in mapping the intracellular signaling network since numerous kinases control the signal transduction process via protein phosphorylation.

1.4   Major strategies in chemical genomics I: Forward chemical genomics

Forward chemical genomics (FCG) starts with probing the genome using small molecule interaction at the cellular level (cell-based approach) (Figure 1.8). Small molecules could activate or inactivate several proteins within a particular pathway or one kind of protein plays multiple roles in the pathway; thus, their overall cellular response can be studied at the cellular and gene expression or protein expression level. Subsequently, or in parallel, the binding partners of the small molecules that were involved in signal initiation can be identified with affinity-based methods.

This approach requires highly specific and tightly binding small molecules for each expressed protein participating in cell cycle pathways and regulation. Novel chemistry would be required, directed at the creation of highly diverse, more complex, small synthetic molecules (libraries), such as natural products. The result of the FCG is a small molecule, which perturbs the signaling pathways resulting in gene expression or phenotypic changes as the cellular response, and in the protein target that the small molecule interacted with. The protein targets of this small molecule's interaction together with the exact function of the proteins within the signaling network can be elucidated.

In chemical genomics screens, the cellular response induced by the small molecules can be measured directly (increased or inhibited cell proliferation, microscopic or visible phenotypic/cell morphological changes)

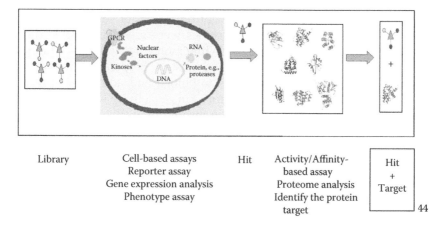

Library	Cell-based assays	Hit	Activity/Affinity-	Hit
	Reporter assay		based assay	+
	Gene expression analysis		Proteome analysis	Target
	Phenotype assay		Identify the protein	
			target	44

Figure 1.8 The general flowchart of forward chemical genetics/genomics.

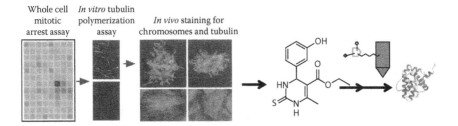

Figure 1.9 Identification of monastrol (an inhibitor of the normal mitotic spindle formation) and its molecular target (a molecular motor protein, kinesin, Eg5) using chemical genomics.

or indirectly (reporter gene assays, gene-expression profiling methods, mRNA, and DNA microarrays) (Figure 1.8).

In a pioneering forward chemical genetic screen (whole-cell mitotic arrest assay detected by fluorescence microscopy), a cell-permeable small molecule, monastrol (Figure 1.9), was identified, as it caused inhibition of the normal mitotic spindle formation but did not affect normal tubulin formation. In subsequent studies, testing the inhibition of the formation of the mutant phenotype led to the identification of the primary molecular target in the signaling cascade, a molecular motor protein, kinesin, Eg5. Monastrol treatment showed a phenotype identical to the blocking of Eg5 function by microinjection of Eg5-specific antibodies (Kapoor et al., 2000).

The forward chemical genetics approach, thus, resulted in a new lead compound, a new therapeutic target, as well as a target-specific chemical probe for cell biology (Mayer et al., 1999). Since cell division has a highly dynamic nature, the temporal control with small molecules over protein functions would be particularly valuable in dissecting biological mechanisms. In further studies, the entire process of cell division was investigated and various small molecules were identified providing a temporal control at different stages. While monastrol trapped cell mitosis in an early stage after removal of monastrol, hesperadin (an Aurora kinase inhibitor) caused the spindles to bipolarize leading to improper chromosome attachment. As the Aurora kinase activity was restored by removal of hesperadin (Figure 1.10) the chromosome attachment was corrected (Lampson and Kapoor, 2006).

1.4.1   Model organisms used in forward chemical genomics

Monitoring the cellular response to small molecules can be efficiently studied by model organisms.

Yeast (*Saccharomyces cerevisiae*) (Shawn et al., 2008) cells are easy to grow; it contains approximately 6000 genes and shows high genetic

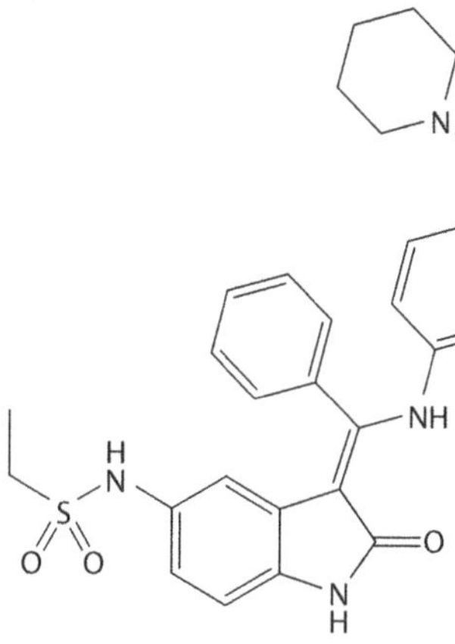

Figure 1.10 Hesperadin, an Aurora kinase inhibitor, which interferes with mitosis.

conservation with humans. Yeast is suitable for using in high-throughput screens.

Large numbers of signaling pathways are very similar between this model organism and mammalian systems, which makes it particularly attractive in chemical genomics screens.

Yeast is often used in synthetic lethality screens, where wild-type strains or strains having modified genomes via mutation or gene deletion are treated with small molecules, and the resulting phenotypic changes could reveal specific alterations and their functions within the intracellular signaling pathways after comparison of the strains.

The zebra fish (*Danio rerio*) (MacRae and Peterson, 2003) is a whole organism suitable as a model for chemical genetics screening. The zebra fish is widely used in developmental biology and genetics and is frequently used for rapid toxicity profiling of small molecules since they are vertebrates with discrete organs such as a brain, sensory organs, heart, muscles, bones, and so forth; and these organ systems are very similar to their human counterparts.

Furthermore, a zebra fish is rather small in its early embryonic stage, able to live in a 200 microL well of a typical 96-well plate. Finally, zebra fish embryos are completely transparent, which allows for monitoring of the processes in every organ leading to characteristic morphological changes upon treatment with small molecules without the need for dissecting or sacrificing the animal.

Other organisms applied in chemical genomics screens include the fruit fly (*Drosophila melanogaster*), and *Caenorhabditis elegans* (*C. elegans*), which is a roundworm.

1.4.2   HT screening techniques applied in forward chemical genomics

1.4.2.1   Synthetic lethal chemical screening

Synthetic lethal chemical screening is basically a combination of classical and chemical genetics and it is particularly suitable for the identification of potential antitumor agents.

The purely genetic origin of synthetic lethality arises when a combination of mutations in two or more genes leads to cell death, whereas a mutation in only one of these genes does not. In a synthetic lethal genetic screen, it is necessary to begin with a mutation that does not kill the cell, although it may confer a phenotype, and then systematically test other mutations engineered to the cell line which leads to lethality. In the synthetic lethal chemical screening, the search for a lethal second mutation is substituted with screening for small molecules that produce cell death. This approach provides a novel tool to discover genotype-selective antitumor agents that become lethal to tumor cells only in the presence of specific oncoproteins or in the absence of specific tumor suppressors (Dolma et al., 2003). Such genotype-selective compounds either target oncoproteins directly or other critical proteins involved in the oncoprotein-linked signaling network. The major advantage of genotype-selective antitumor agents is that they are active primarily in the presence of specific genetic elements. Such a method is also useful for dissecting the oncogenic signaling network and tailoring chemotherapy to specific mutated tumor types. In order to characterize the lethality of the small molecules in tumorigenic cells, the compounds were also screened for cell viability against isogenic primary cells as a reference cell line.

A *tumor selectivity score* was introduced for each compound by dividing the IC_{50} value for the compound in the parental, primary BJ cells by the IC_{50} value for the compound in engineered tumorigenic cells. Apart from confirming the antitumor effects of various known chemotherapic agents such as doxorubicin, daunorubicin, mitoxantrone, camptothecin, sangivamycin, echinomycin, and bouvardin, synthetic lethal chemical screening of a diverse combinatorial library (containing 23,500 compounds) led to identification of a novel compound, Erastin (Figure 1.11), that selectively kills cells expressing the small T oncoprotein and oncogenic RAS.

Erastin exhibits greater lethality in human tumor cells harboring mutations in the oncogenes HRAS, KRAS, or BRAF. Using affinity purification of the amino-alkyl derivative (Erastin A6) immobilized to the resin and mass spectrometry, it was discovered that Erastin acts through mitochondrial voltage-dependent anion channels (VDACs)—a novel target for anticancer drugs (Yagoda et al., 2007). In further studies, 47,000 compounds were screened against the RAS oncogene-containing tumorigenic cell lines and normal cell lines (Yang and Stockwell, 2008). While RSL5

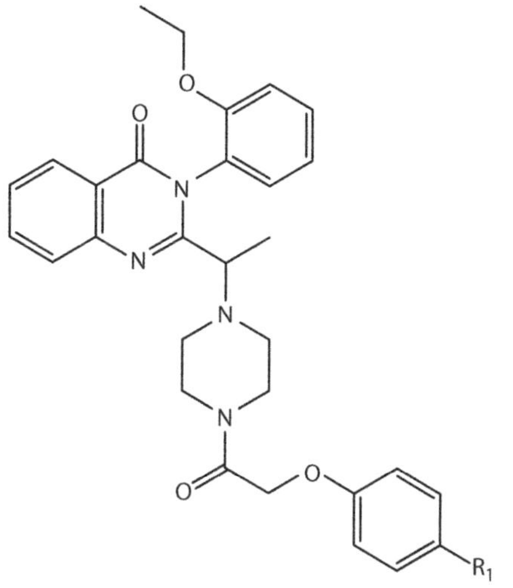

Erastin R$_1$ = Cl
Erastin A6 R$_1$ = CH$_2$NH$_2$

Figure 1.11 Erastin, a potential genotype-specific antitumor agent and its anchorable analog (Erastin A6).

acts through VDACs to activate the RAS oncogenic pathway, RSL3 is activated in a similar death mechanism but in a VDAC-independent manner. It was found that cells transformed with oncogenic RAS have increased iron content relative to their normal cell counterparts through up-regulation of transferrin receptor 1 and down-regulation of ferritin heavy chain 1 and ferritin light chain (Figure 1.12).

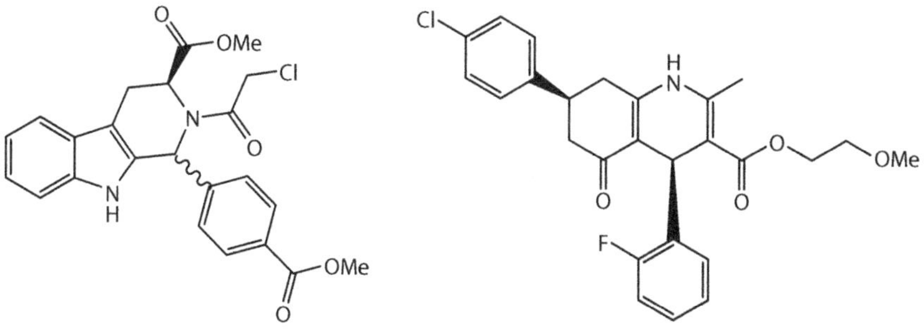

RSL3

RSL6

Figure 1.12 Second-generation antitumor drug candidates having novel chemotypes, which act on mitochondrial voltage-dependent anion channels.

1.4.2.2   Cytoblot assays

A cytoblot assay is a miniaturized cell-based assay format that interrogates the cellular pathways in mammalian systems using small molecules. This whole-cell immunodetection assay can sensitively detect changes in specific cellular macromolecules in high-density arrays of mammalian cells. Cytoblotting can be used to monitor biosynthetic processes such as DNA synthesis, and posttranslational processes such as acetylation and phosphorylation.

In the cytoblot assay, double immunochemical detection was applied: the primary antibody detects, for example, phosphorylated proteins while a second antibody linked to horseradish peroxidase provides a chemi-luminescence readout (Stockwell, Haggarty, and Schreiber, 1999). The applicability of the cytoblot assay was demonstrated by natural product screening through the identification of marine sponge extracts exhibiting genotype-specific inhibition of 5-bromodeoxyuridine incorporation and suppression of the antiproliferative effect of rapamycin. In another application, a similar setup was used to identify the phosphorylation pattern of multiple signaling proteins in response to known kinase inhibitors (Chen et al., 2005).

1.4.2.3   Reporter gene assays

A reporter gene is a gene that is attached to a regulatory sequence of another gene of interest in cell culture, animals, or plants. Reporter genes are in general used as an indication of whether a certain gene has been taken up by or expressed in the cell or organism population.

Stockwell investigated TGF-β signaling pathways by constructing a reporter gene assay using a p3TPLux reporter gene (Stockwell et al., 1999). The extent of reporter gene activation was monitored by standard luciferase chemiluminescence readout. In the primary screen, a 16,000-membered diverse compound library was tested for activating the reporter gene p3TPLux8 in a stably transfected mink lung epithelial cell line. The most active compounds (Figure 1.13) in the reporter screen were used for gene expression studies in the model organism *S. cerevisiae*, whose genome was available in cDNA microarray (5800 altogether). The compound reproducibly increased the expression of five genes (*HSP26, ZRT1, FET3, YDR534C,* and *YOL155C*) more than twofold. The function of the

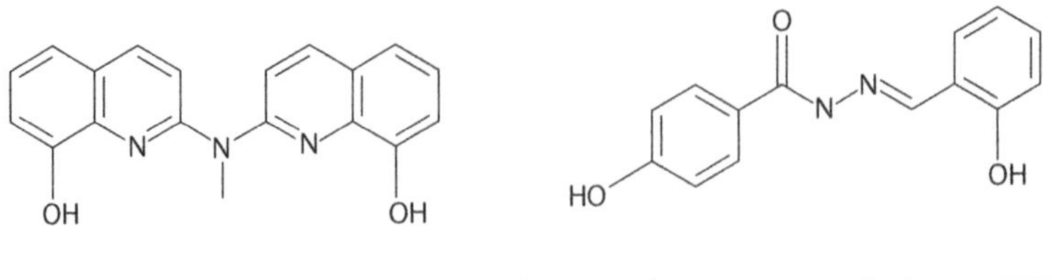

Figure 1.13 Compounds that activate the transforming growth factor (TGF)-β-responsive reporter gene.

first three was already known, and all were involved in metal ion (copper, iron, or zinc) homeostasis.

1.4.3   Case studies in forward chemical genomics

Reversing the phenotypic changes induced by viral oncogenes

There are chemical genetic examples for the reversal of diseased pheno-types by addition of small molecules. The viral oncogenes *v-ras* and *v-src* transform fibroblast cells from the NIH3T3 cell lines to a characteristic spindly morphology, which is considered a pathological state. By the administration of the natural product depudecin (Figure 1.14), this state can be reverted into the wild-type, original, flattened morphology.

Thus, potential therapeutic agents could suppress the formation of a diseased state conditionally. Since the primary target for depudecin is his-tone deacetylase, it confirms that this is an excellent druggable target for anticancer therapy (Kwon et al., 1998).

Identification of small molecules promoting specific differentiation of embryonic stem cells (Sachinidis et al., 2008)

Cell replacement therapy for severe degenerative diseases such as diabe-tes, myocardial infarction, or Parkinson's disease through transplantation of somatic cells generated from embryonic stem (ES) cells is a very prom-ising new approach.

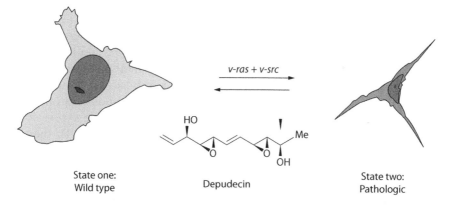

Figure 1.14 Depudecin reverts the phenotypic changes induced by viral onco-genes. (From Kwon, H. J. et al., 1998, Depudecin Induces Morphological Reversion of Transformed Fibroblasts via the Inhibition of Histone Deacetylase, *Proc. Natl. Acad. Sci. USA*, 95:3356–61.)

ES cells harvested from the inner cell mass (ICM) of the early embryo can proliferate indefinitely *in vitro* while retaining the ability to differentiate into all somatic cells, thereby providing an unlimited renewable source. Identifying synthetic small molecules and signal cascades involved in specific organ cell differentiation processes toward a defined tissue-specific cell type are extremely important for optimizing the generation of somatic cells *in vitro* for potential therapeutic use. Forward chemical genetic embryonic stem cell models offer a viable approach to identifying small molecules promoting specific differentiation of ES cells to the desired cell phenotype. Cardiomyogenesis was chosen to demonstrate the utility of the chemical genetic screening of small molecule libraries to identify potential agents capable of enhancing differentiation toward cardiomyocytes.

After culturing of transgenic ES cells in multiwell plates, a library of small molecules was screened by addition of one single compound per well. After differentiation, small molecules can be identified participating in the generation of the desired phenotype. Interestingly, verapamil (a class IV antiarrhythmic agent, an L-Type Ca2+ channel blocker) was identified exhibiting a striking procardiomyogenic effect.

Identification of AML1/ETO oncoprotein modulators (Corsello et al., 2009)

Approximately 15% of all acute myeloid leukemia (AML) cases possess t(8;21) chromosome translocation that gives rise to the AML1/ETO fusion oncoprotein. AML1/ETO consists of the N-terminal DNA-binding domain of AML1, a transcription factor essential for definitive hematopoiesis, and almost all of ETO, a protein thought to function as a corepressor for a variety of transcription factors. AML1/ETO impacts multiple processes and interferes with multiple signal transduction pathways, promoting leukemogenesis. To identify AML1/ETO modulators, a small molecule library containing 2480 FDA-approved drugs and known bioactive compounds was screened using a gene expression-based chemical genomic approach. Gene expression signatures were used as surrogates for the expression versus loss of the translocation in AML1/ETO-expressing cells, therefore it was first necessary to define a signature for AML1/ETO abrogation, which was done by RNAi-mediated knockdown of AML1/ETO. The final signature constituted 28 genes. After 72 hours of incubation, the cells (derived from AML1/ETO positive AML) were lysed, and cDNA synthesis was performed. The expression level for each of the 28 genes in the signature was detected by a bead-based fluorescence system. Even though transcription factors were historically considered to be various undruggable compounds that belonged to corticosteroids and dihydrofolate reductase

(DHFR) inhibitor classes (e.g., methotrexate and methylprednisolone), they induced the AML1/ETO abrogation signature at low nanomolar concentration. Hit confirmation was done by cell-based assays, where the compounds induced myeloid differentiation, dramatically reduced cell viability, and ultimately induced apoptosis.

Discovery of new small molecules inhibiting angiogenesis (Kwon, 2006)

Angiogenesis is a multistep process of new blood vessel formation by endothelial cells, which is critical for normal physiology such as embryonic development and wound healing. In a cell-based screening for new angiogenesis inhibitors, a library of small molecules consisting of natural products (including plant extracts) and synthetic compounds were treated to endothelial cells. A methanol extract of *Curcuma longa* showed potent inhibition of endothelial cell proliferation as well as tube formation, one of distinct differentiation phenotypes of endothelial cells. Based on these results, a curcumin-based focused library was constructed and HBC (Figure 1.15) showed potent inhibitory activities against the proliferation of several human cancer cells.

HBC was then investigated for the identification of their functional target using known angiogenesis-related targets, such as aminopeptidase N (APN). Since HBC did not inhibit APN, an entirely different mechanism is thought to be responsible for the inhibition of the tumor cell growth. Since the biological activity of HBC was in the 10 micromolar range, it can be expected that HBC has most likely low affinity to its putative target. Since affinity chromatography isolation of the protein requires high-affinity binding, a phage display biopanning analysis was applied which is widely used for target identification of low-affinity ligands. T7 phages

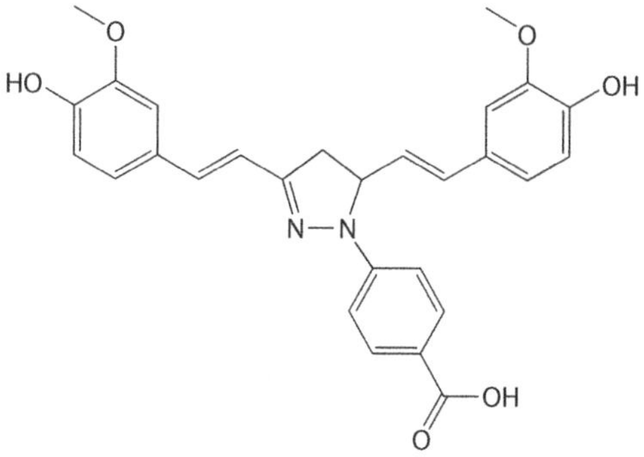

Figure 1.15 HBC, a curcumin analog, is a potent inhibitor of angiogenesis and of proliferation of several human cancer cells.

encoding human cDNA library led to binding phage amplification using HBC affinity matrix. After isolation, amplification of each phage clones DNA sequencing revealed that the molecular target is Ca^{2+}/calmodulin (Ca^{2+}/CaM) as a direct target protein of HBC. Ca^{2+}/CaM is a calcium binding protein, which is implicated in a variety of cellular functions, including cell growth and proliferation.

Identification of an apoptosis inducer with a combined cell-based and caspase-activity screen (Cai, Drewe, and Kasibhatla, 2006)

A cell-based phenotypic assay was developed and used for the identification of small molecules that exhibit apoptosis-inducing activities. The assay was also linked to measure caspase activity which is a reporter of the distinct apoptosis process. The mechanism of apoptosis involves a cascade of sequentially activated initiator and effector caspases that are a family of cysteine proteases which require aspartic acid residues for cleavage. Among these caspases, caspase 3, 6, and 7 are key effectors that cleave multiple protein substrates, such as poly(ADP)ribose polymerase (PARP) in cells, leading irreversibly to cell death. Since inadequate apoptosis is known to play a critical role in the development of cancer, promoting apoptosis is a potential strategy for cancer therapy. A cell- and caspase-based high-throughput screening (HTS) system was developed to identify small molecules from commercial libraries that perturb the apoptosis signal pathway, induce apoptosis, and activate caspase-3/7. This assay technology also monitors the activation of downstream caspase-3/7 using a fluorescent caspase-3/7 substrate. Human breast cancer T47D cells in 96- or 384-well microtiter plates containing 10 µM of test compound were incubated for 24 h for the induction of apoptosis. Fluorescent caspase-3/7 substrate (N-(Ac-DEVD)-N'-ethoxycarbonyl-R110) was then added to the cells and incubated for 3 h. Compounds that induce apoptosis and activate the caspases yield a fluorescent signal higher than the background (hits were considered as having a threefold increase in fluorescence). Several active compounds of different classes were found in the combined assay and the putative target was identified by photoaffinity labeling: the tail interacting protein 47 (TIP47) and insulin-like growth factor II (IGF II) for aryloxadiazoles; transferrin receptors for Gambogic acid (Figure 1.16).

Restoring the function of the tumor suppressor with small molecules

When a tumor suppressor gene has been deleted from the genome of a tumor cell, it is rather challenging to find a small molecule to restore the function of the encoded tumor suppressor protein. Kau et al. (2003)

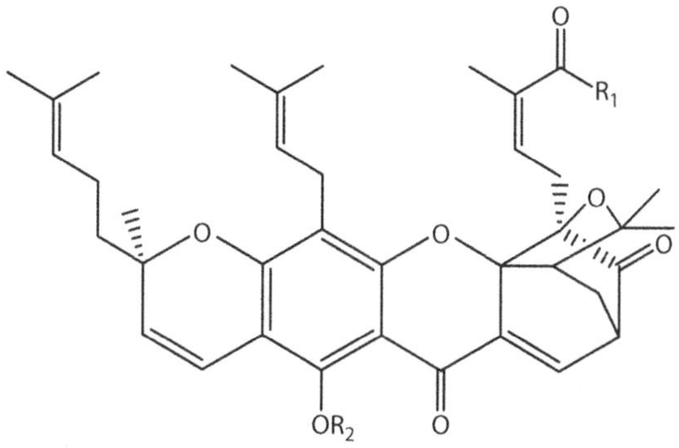

Figure 1.16 Gambogic acid as a potent apoptosis inducer.

demonstrated that it is possible to restore at least one function of the tumor suppressor PTEN using small organic molecules in cells that lack PTEN (Kau et al., 2003). The PI3K/PTEN/Akt signal transduction pathway plays a key role in many tumors. Downstream targets of this pathway include the Forkhead family of transcription factors (FOXO1a, FOXO3a, and FOXO4). PTEN functions as a lipid phosphatase upstream of Akt, a kinase that phosphorylates FOXO1a and thereby regulates its nuclear localization. In PTEN null cells, FOXO1a is inactivated by PI3K-dependent phosphorylation and mislocalization to the cytoplasm. A high-through-put chemical genetic screen (involving an 18,000 compound library) was performed for small molecules that could localize FOXO1a to the nuclei of PTEN-deficient cells. The screen revealed both compounds that were PTEN/Akt/FOXO1a pathway-specific and compounds that were general nuclear export inhibitors acting on nuclear export receptor CRM1. Analysis of the chemical structures having restoring activity revealed that they contain thiol-reactive electrophilic functionalities, such as Michael acceptors (such as α,β-unsaturated carbonyls) (Figure 1.17). They covalently bind to cysteine 528 of CRM1.

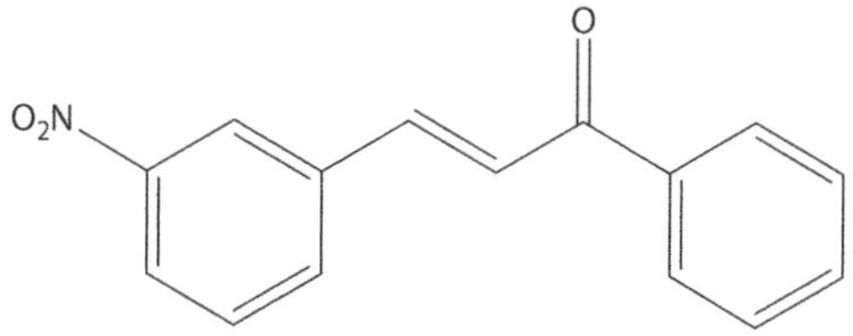

Figure 1.17 A typical α,β-unsaturated carbonyl compound that restores the function of a tumor suppressor.

1.4.4   Biology

1.4.4.1   Combination chemical genetics

While the primary objective of chemical genomics is to find a selective modulator (perturbant) of a protein target and the consequences of the single small molecule–protein interactions in attempts to associate them with alterations in signaling pathways, combination chemical genetics (Lehár et al., 2008) attempts to model intracellular networks in their real complexity. In fact, a single mammalian cell typically orchestrates the activities of over 20,000 genes, 100,000 proteins, and thousands of small molecule lipids, carbohydrates, and metabolites, each of which may be expressed at differing levels over time. Although single perturbations are effective at determining which components in a system are essential for a phenotype, functional connections between components are better characterized through combination effects. Multiple molecules within the cell could cause a combination effect which can be either synergy or antagonism. Different small molecules most likely hit different protein targets in the signaling network which could result in a synergetic effect. This concept is similar to analyzing the effects of multiple mutations applied in classical genetics (Figure 1.18).

Combination drugs and multicomponent therapeutics (e.g., herbs and natural products) have been used since ancient times. There is widespread evidence that combinations of compounds can be more effective than the sum of the effectiveness of the individual agents themselves, a result that can be rationalized using principles derived from modern systems and molecular biology. Combination chemical genetics coupled with the informatics methods of system biology could provide deeper insights into the

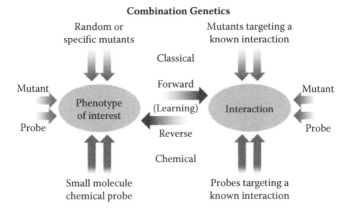

Figure 1.18 Comparison of classical combination genetics and chemical combination genetics. (From Lehár, J. et al., 2007, Chemical Combination Effects Predict Connectivity in Biological Systems, *Mol. Syst. Biol.*, 3:80, doi:10.1038/msb4100116.)

combination effects of small molecules while interacting with various proteins in the signaling network and could lead to more effective drugs with reduced side effects (Curtis, Borisy, and Stockwell, 2005).

Combination chemical genetics allows synergy mapping, which was carried out in an anticancer screen in HCT116 cells testing all pairs of 90 drugs (including various kinase inhibitors) (Lehár et al., 2007). The responses to pairings of kinase inhibitors provided support for an association between combination effects and connectivity. Combinations between drugs targeting the phosphatidylinositol-3 kinase (PI_3K), protein kinase C (PKC), or mitogen-activated protein kinase (MAPK) signaling pathways showed exceptionally strong synergies and correlations, which corresponds to the hypothesis that HCT116 cell proliferation is closely associated with epidermal growth factor receptor (EGFR)-mediated signaling. These findings are in accordance with the novel multitarget, anticancer drug discovery concept, which attempts to control multiple pathways with one single molecule (Petrelli and Giordano, 2008). Thus, multitarget or combination approaches offer improved control by perturbing multiple components of the disease network (Csermely, Agoston, and Pongor, 2005).

1.4.5   Toxicogenomics: Reference libraries of toxic compounds

Toxicogenomics is concerned with the identification of potential human and environmental toxicants, and their putative mechanisms of action, through the use of genomics resources. Its goal is to find correlations between toxic responses to toxicants and changes in the gene expression profiles of the objects exposed to such toxicants (Waters, Stasiewicz, and Merrick, 2008). The changes in gene expression induced by the test agents in the model systems are analyzed, and the common set of changes unique to that class of toxicants, termed a *toxicant signature*, is determined (Khor, Ibrahim, and Kong, 2006). This signature is derived by ranking across all experiments the gene expression data based on relative fold induction or suppression of genes in treated samples versus untreated controls and selecting the most consistently different signals across the sample set. A different signature may be established for each prototypic toxicant class. Once the signatures are determined, gene expression profiles induced by unknown agents in the same model systems can then be compared with the established signatures. A match assigns a putative mechanism of action to the test compound (Figure 1.19).

Early toxicogenomic profiling of drug candidates gained much importance in early drug discovery. It could reduce the attrition rate through early identifying of possible side effects (e.g., hepatotoxicity) and such compounds can be filtered out from the discovery pipeline.

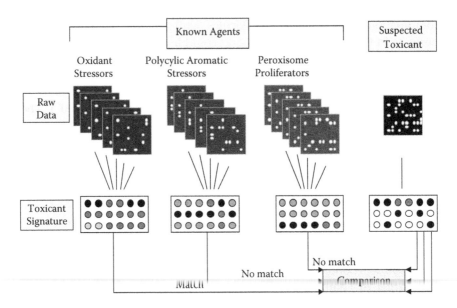

Figure 1.19 Toxicogenomics strategy to identify potential toxicants based on gene expression fingerprints.

1.4.6 Drug chemical genomics

Drug resistance can be considered one of the major challenges in cancer therapy. Monitoring the changes of the expression profile of the cells coming from the different stages of cancer would reflect the progression of the disease as well as the resistance or retained sensitivity toward chemotherapeutic agents (Shingo et al., 2002). Treating cancer cell lines *in vitro* with antitumor agents could provide valuable information about the prospects for the treatment. The evolving drug resistance and the sensitivity can be associated with a particular expression pattern, thus, coadministration of multidrug resistance-reversing agents with the cytotoxic drug could recover the original expression profile and the antitumor effectivity.

In a recent study, comparison of the expression profiles in lymphoid cancer resistant to glucocorticoid treatment and screening for adjuvants revealed that, interestingly, the addition of rapamycin (an immunosuppressant agent) could recover the expression pattern of the glucocorticoid sensitive cells and, in fact, the efficiency of the treatment (Guo et al., 2006). The practice of drug chemical genomics could contribute to the evolution of personalized medicine monitoring the inherent or altered sensitivity toward a particular agent or treatment. In the future, expression profiles as well as unique genetic biomarkers could be routinely used in diagnostics (Peták et al., 2010) as well as treatment and contribute to personalized treatment in oncology (Gonzalez-Angulo, Hennessy, and Mills, 2010) leading to an increased survival rate.

1.4.7   Identification of protein targets in forward chemical genomics studies

1.4.7.1   HT affinity chromatography approaches

An integral part of the forward chemical genomics strategy is the identification of the target protein that the small molecules interact with and to initiate a perturbance of the signaling network resulting in a cellular response. The primary method for isolation and identification of the protein targets is the pull-down assay, which is based on the classical concept of affinity chromatography (Figure 1.20). The cell-based small molecule hit is appended with a linker (typically an alkyl or polyethylene glycol chain) having a terminal functional group for efficient covalent anchoring on a solid matrix (resin). Letting through the cell lysate on the column packed with affinity resin, the solubilized mixture of proteins elutes according to their binding strength.

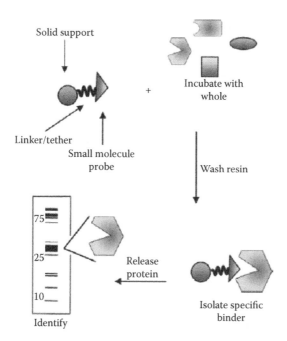

Solid support

Incubate with whole

Linker/tether

Small molecule probe

Wash resin

75

25

10

Release protein

Identify

Isolate specific binder

Figure 1.20 The major elements of the pull-down assay. (From Terstappen, G. C. et al., 2007, Target Deconvolution Strategies in Drug Discovery, *Nat. Rev. Drug Disc.*, 6:891–903.)

The collected protein fractions are digested and the amino acid sequence of the resulting peptide fragments are determined by tandem mass spectrometry (MS/MS) analysis. The exact mass of the fragments is matched to peptide sequences using large databases which enables the identification of the whole-length protein. The main challenge of the pull-down assay is to verify the specificity of the binding as well as the optimal introduction of the linker to the parent compound (site of attachment, length of the linker, etc.) in order to retain the activity and specificity (Terstappen et al., 2007) (Figure 1.20).

There are various strategies for separating the specific binding proteins from the nonspecific interactions including the application of a close but inactive immobilized analog in a separate experiment (inactive comparison column) or using the parent nonimmobilized ligand as a competitor in large excess in the pull-down assay of the immobilized active ligand. The specificity is verified if, in both cases, the specific binding protein cannot be detected due to either the lack of binding activity (inactive column) or the prior saturation of the binding sites, which eliminates the protein binding to the active column.

The introduction site of the linker can be assessed by available structure-activity relationship date or, in the absence of such information, the linker can be attached at various regions of the active ligand in a combinatorial manner (Darvas et al., 2001). One of the widespread applications of the pull-down assay is the affinity profiling of multitarget inhibitors of large target families in various cells and tissues. The approach also enables the comparison of the expression level of the various subtypes (isozymes) within the target families concerned. Since immobilized small molecule inhibitor libraries are used to probe the functional proteome in various sources, this methodology is often termed *chemical proteomics*.

1.5 Major strategies in chemical genomics II: Reverse chemical genetics/genomics

Reverse chemical genetics (RCG) starts with an identified novel protein target and screens for small molecules that specifically bind to the protein (Figure 1.21). It looks similar to the target-based drug discovery although in this case the function of the binding protein is not or not completely understood. The major objective is then to investigate whether the target-specific small molecule could cause a phenotypic or gene expression change in a cell-based assay, which helps in identifying the intracellular signaling process that implicates the protein concerned. This approach is analogous to reverse genomics where intentional, well-designed mutations are introduced or genes are knocked down in order to induce phenotypic changes and to study the intracellular processes. In RCG, the initial

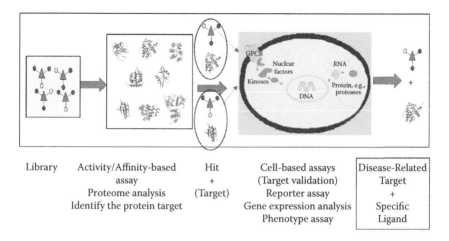

Library   Activity/Affinity-based   Hit   Cell-based assays   Disease-Related
assay   +   (Target validation)   Target
Proteome analysis   (Target)   Reporter assay   +
Identify the protein target   Gene expression analysis   Specific
Phenotype assay   Ligand

Figure 1.21 The general flowchart of reverse chemical genetics/genomics.

target–ligand pairs can be linked to specific changes in the cell behavior in disease-associated assays.

The first step is identifying small molecule interaction in a genome-(proteome-) wide manner using activity/affinity-based approaches, while the second step is a target validation in whole-cell assays. In the initial chemical proteomics step, affinity-based chemical probes can reveal binding affinity signature to proteins within the cell by investigating binding at the proteome level.

1.5.1 Case studies for reverse chemical genomics applications

1.5.1.1 Target validation of purvanolol binding protein

Purvalanol A and one of its close analogs (Figure 1.22a) were selected as the most potent and selective inhibitor against human CDK2 (cyclin-dependent kinase 2) from a large library. In order to determine the cellular effects of the selective inhibitors and confirm the cyclin-dependent kinase (CDK)-directed cell cycle inhibition, which is their therapeutically exploitable value, purvalanol A (R = H) was tested in 60 human cancer cell lines for tumor growth inhibitory activity, as well as for effects on mRNA levels of yeast genes. For comparison, a weak analogous inhibitor (R = Me) was also involved in the studies. Overall, 6200 genes were monitored, and 194 showed greater than twofold change in transcript level for purvalanol A and a much smaller number for the weak inhibitor reference. Among the down-regulated transcripts, five were directly associated with the cell cycle, confirming the function of the target (Gray et al., 1998).

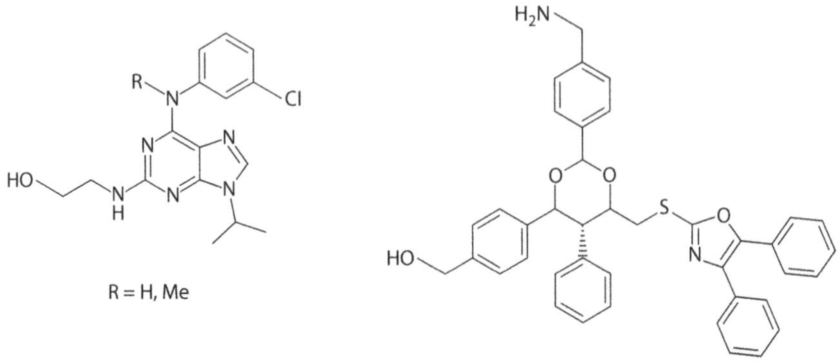

Figure 1.22 Purvalanol analogs, a potent CDK2 inhibitor (a), and uretupamine A, which binds to Ure2p, a protein involved in glucose signaling (b).

1.5.1.2 Dissecting glucose signaling pathways

A yeast regulatory protein, Ure2p, that plays a role in repressing certain genes and transcription factors (PUT1, PUT2, PRB1, Gln3p, and Nil1p) was intended to be studied with small molecule modulators. However, binding partners, especially modulators, were not known; thus the first small molecules were identified from a 3600-member library by a microarray-based fluorescent binding assay (Kuruvilla et al., 2002) that showed affinity to the protein. One of the most active compounds (7.5 microM), uretupamine A (Figure 1. 22b), was tested for modulation of Ure2p in a cell-based reporter assay, indicating increase of the expression of PUT1, confirming that its expression is under the control of Ure2p. In a DNA array study, transcription profiling revealed that treatment with uretupamine A resulted in the up-regulation of five Ure2p-dependent genes (PUT1, PUT2, PRB1, NIL1, and UGA1) as expected by blocking the regulatory protein. Since the uretupamine A-induced genes are among those normally up-regulated when glucose is removed from the media, it is suggested that Ure2p is one of the downstream effectors of the glucose-sensitive signaling and can be blocked through that protein. These studies are essential to understanding the disease mechanism of type II diabetes.

1.5.2 Surrogate compounds and molecular probes in reverse chemical genetics

Target-specific, small molecules (Potuzak, Moilanen, and Tan, 2004) are used as *surrogate* ligands in cell-based assays for identifying or validating the function of the target in the signaling network through phenotypic change or alteration in the gene expression profile. Such surrogate ligands

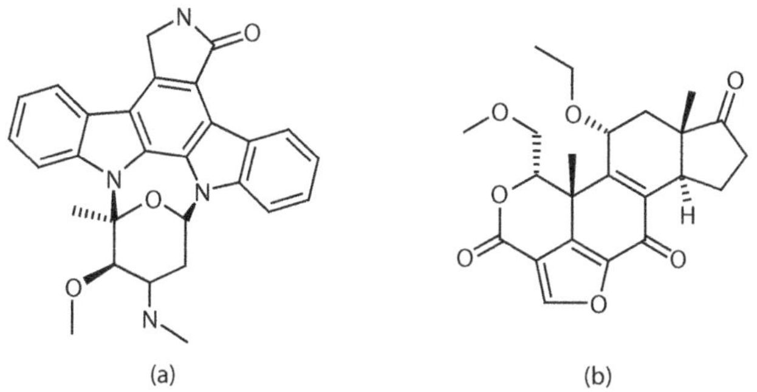

(a)   (b)

Figure 1.23 The structure of staurosporine (a), as a robust, nonselective kinase inhibitor, and wortmannin (b), a highly specific phosphoinositide 3-kinase inhibitor.

are particularly important for broad, highly conserved gene families, like protein kinases, that contain more than 500 closely related targets and constitute 2% of the human genome. Kinases transmit the intracellular signals through protein phosphorylation and provide druggable targets for the therapy of cancer, inflammation, metabolic or autoimmune diseases, and so forth. To develop specific binders for each gene product in such families is particularly challenging due to the highly conserved amino acid sequences. Only a few examples are demonstrated for specific kinase inhibition (PI3K, p38 kinase, MEK), while the rest are broad-spectrum ATP-mimicking inhibitors, for example, staurosporine (Figure 1.23a). An unprecedented example for *surrogate* compounds is wortmannin (Figure 1.23b), a unique phosphoinositide 3-kinase inhibitor (PI-3K), that irreversibly interacts with the enzyme with exceptional selectivity (>3000-fold) and high affinity (<<10 nm).

1.5.3   Orthogonal chemical genetics

In the absence of target-specific modulator small molecules, Shokat and coworkers (2001) developed a method described as the *bump-and-hole* approach for orthogonally investigating one single protein in the target family and the target's role within the signaling network. They introduced a bulky group in a natural, nonspecific ATP analog substrate (Figure 1.24), and a functionally silent mutation was inserted into the active site of a protein kinase, creating an enlarged cavity. A kinase with such a cavity is also called *analog-sensitive kinase alleles* (ASKAs) (Witucki et al., 2002).

The engineered enzyme was only able to interact with the bulky substrate that enables differentiation within the target family. In light of this fact, genetic engineering and chemistry can be brought together to

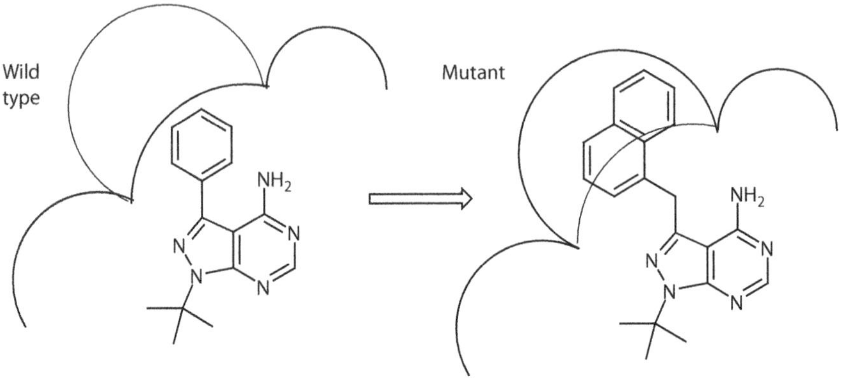

Figure 1.24 Orthogonal chemical genetics: development of analog-sensitive kinase alleles (ASKAs).

design orthogonal protein ligand pairs, which enable investigation of (activate or inactivate) signaling pathways associated with the kinase target concerned separately from the other kinases. The complementary protein–ligand pairs provide unique opportunities for identifying selective inhibitors of the downstream signaling in cell-based assays to be developed. Orthogonal protein–ligand pairs can also be used to validate multifunctional tumor (kinase) targets by identifying the primary pathways that are responsible for the cytotoxic activity (Burkard and Jallepalli, 2010) (Figure 1.24).

1.6   Chemical proteomics

Chemical proteomics is a parallel investigation of the interactions between small molecule libraries and their binding proteins at the proteome level.

1.6.1   Drug chemical proteomics

There are huge target families that represent a particular challenge to developing selective modulators such as inhibitors for protein kinases (Fedorov et al., 2007).

Multitarget or robust kinase inhibitors with a known activity profile (Fabian et al., 2005) can be used to detect the expression level of different kinase targets through affinity chromatography (*pull-down* assay). Binding profile analysis of the various immobilized inhibitors (Daub et al., 2004) would allow comparing the expression level of various kinases in different tissues or cells, or upon treatment with small molecule perturbants in chemical genomics. Figure 1.25 shows the attachment points of the kinase inhibitors to the solid support involved in the study (Figures 1.25 and 1.26).

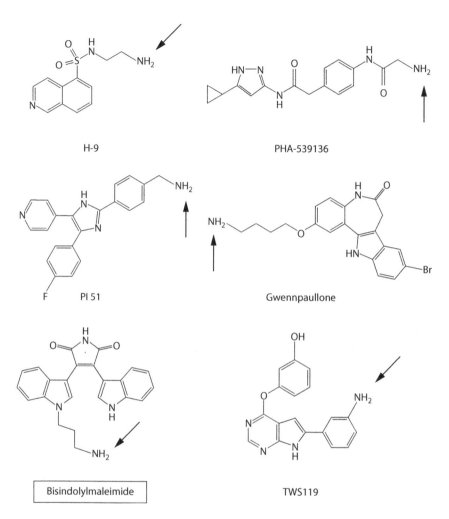

Figure 1.25 Various multitarget kinase inhibitors for chemical proteomics indicating the attachment point to the affinity matrix. (From Daub, H. et al., 2004, Evaluation of Kinase Inhibitor Selectivity by Chemical Proteomics, *Assay Drug Dev. Technol.*, 2(2):215–24.)

Comparison of the different target affinity profiles could reveal the mechanism of the action, possible side effects, target, antitarget, or off-target activities (Rix and Superti-Furga, 2009). Comparison of three anti-leukemia drugs (imatinib, nilotinib, and dasatinib) that inhibit primarily the oncogenic BCR-ABL kinase (Hantschel, Rix, and Superti-Furga, 2008) revealed that dasatinib (a second-generation therapeutic) has a significantly different profile including more than 30 additional kinases in chronic myeloid leukemia (CML) cells (Figure 1.26). Among the most

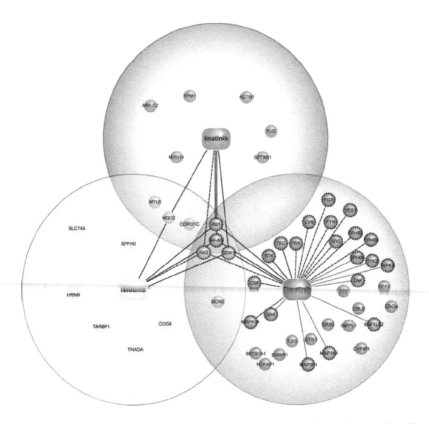

Figure 1.26 Kinase activity profile of three antileukemia drugs (imatinib, nilo-tinib, and dasatinib) that primarily inhibit the oncogenic BCR-ABL kinase. (From Rix, U. and Superti-Furga, G., 2009, Target Profiling of Small Molecules by Chemical Proteomics, *Nat. Chem. Biol.*, 5(9):616–24.)

prominent targets of dasatinib were TEC family kinases BMX, TEC, and BTK, which are particularly important proteins due to the crucial roles in the T- and B-cell receptor signaling and development. This explains that dasatinib shows increased potency and ability to inhibit the majority of imatinib-resistant mutations.

In combination with forward chemical genomics, drug proteomics investigates the cellular response upon small molecules treatment with affinity-based methods: the alteration in the signal transduction pathways and particularly in the kinome (a subset of the genome consisting of the protein kinase genes) induced by the small molecules through binding to the intracellular targets. The resulting expression (binding) profile changes in the kinome can be monitored by immobilized kinase inhibitors. Such information is highly valuable in the elucidation of the mechanism of action for drugs. The alteration of the kinome upon treatment

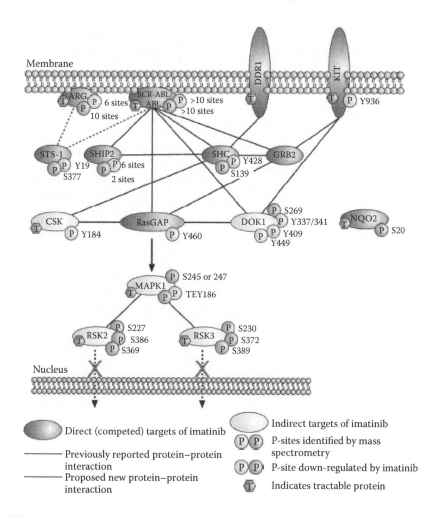

Figure 1.27 Direct and indirect targets of imatinib within the signaling network. (From Bantscheff, M., Eberhard, D., Abraham, Y., Bastuck, S., Boesche, M., Hobson, S. et al., 2007, Quantitative Chemical Proteomics Reveals Mechanisms of Action of Clinical ABL Kinase Inhibitors, *Nat. Biotechnol.*, 25(9):1035–44.)

with imatinib (a BCR-ABL kinase inhibitor against CML) was investigated with seven anchored nonselective kinase inhibitors (Bantscheff et al., 2007) (affinity separation followed by MS/MS analysis). The affinity profiling was complemented with a phosphorylation analysis of the proteins involved in the kinase signaling network which revealed downstream effects. All the changes were summarized in a signal interaction map (Figure 1.27) displaying the direct targets for imatinib within the network as well as the indirect targets where the phosphorylation was reduced.

This confirmed that imatinib inhibited the signal transduction in that particular pathway. This novel integrated concept is also called *Chemical Pathway Proteomics* (Kruse et al., 2008).

1.6.2   Chemical microarrays for studying protein–ligand interactions

One of the most promising tools for studying protein–ligand interactions in a proteome-wide and high-throughput manner is the use of chemical microarrays, which are comprised of thousands of small molecules attached to a solid surface in an ordered format. (MacBeath and Schreiber, 2000; Uttamchandani, Wang, and Yao, 2006; Duffner, Clemons, and Koehler, 2007). Chemical microarrays provide uniform, high-density, and miniaturized protein binding assays. Derivatization chemistries of the solid supports of chemical microarrays were mainly adapted from DNA-microarray technology. Small molecules are attached to the surface through covalent binding of different groups (–OH, –SH, or NH2)—positioned at the end of the linker molecules—by reacting with the active groups of the surface. Because the accessibility of small molecules to their possible interacting partners is crucial in protein–ligand interaction, longer linker arms are preferred.

Freundlieb and Gamer (2002) reported novel technologies for the production of chemical microarrays. Glass plates of microtiter plate format were coated with a gold layer, on which the attachment of the ligands is mediated by an organic film, constituting a self-assembling monolayer. Small organic compounds attached to a linker are immobilized on the gold-layered chip surface. Surface plasmon resonance was used to detect specific interactions between different ligand–receptor, antigen–antibody pairs. This approach was used in custom chemical microarray production for rapid affinity fingerprinting of the S1 pocket of factor VII (Dickopf et al., 2004).

There are two ways to generate a collection of ligand molecules for immobilization on surfaces: chemical library synthesis on micro- or macrobeads and diversity-oriented parallel synthesis of tethered unique compounds (Darvas et al., 2004).

Both approaches utilize combinatorial synthesis, but in the second approach, the tethered molecules are independently synthesized, separated, purified, and analyzed. As an example for the first approach, MacBeath reported a solid-phase combinatorial synthesis combined with thiol-reactive surfaces (Figure 1.28) (MacBeath, Koehler, and Schreiber, 1999).

After completing the synthesis, the beads were separated and distributed by using a *bead-arrayer* in a 384-well format where one bead is distributed to one well. Then, compounds reflecting the one-bead, one-compound

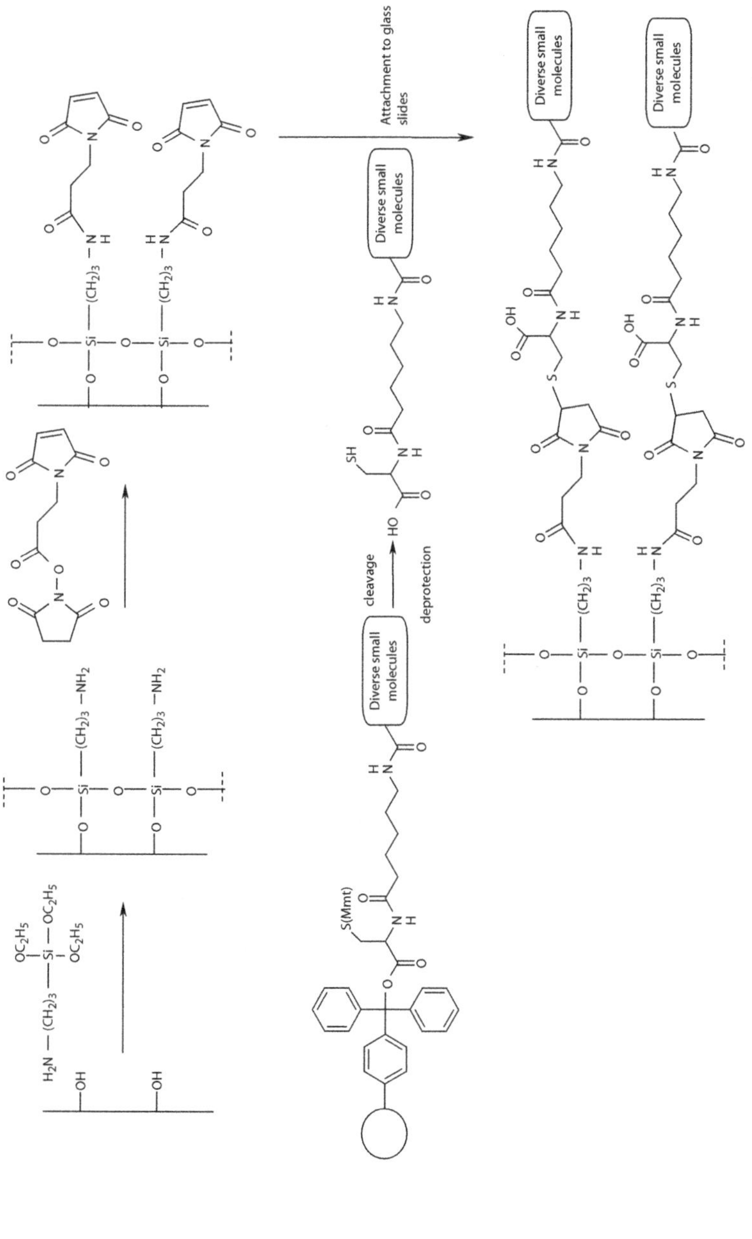

Figure 1.28 Combination of solid-phase synthesis, cleavage, and microprinting to glass slides.

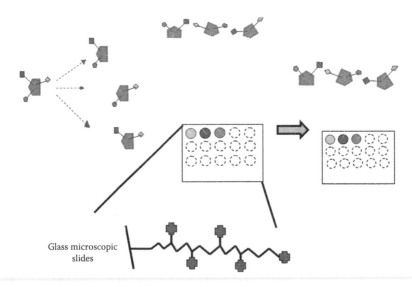

Figure 1.29 Chemical microarray development using high-density reactive glass slides and combinatorially tethered small molecule libraries for immobilization.

concept were cleaved from the beads and then resuspended in a suitable solvent in individual wells and used for spotting (microprinting) the compounds on the surface (Vegas, Fuller, and Koehler, 2008).

In order to prepare proteins and small molecules for microspotting, the compounds should be attached with a spacer or tether, which is long enough to allow undisturbed binding interaction with the proteins. This tagged library approach adds another degree of functionality to the molecule, enabling immobilization or linking reactive or reporter groups (Mitsopoulos, Walsh, and Chang, 2004). Hackler and coworkers (2003) developed a branched dendrimeric spacer system on glass slides for preparing chemical microarrays, which allows high loading capacity of the ligands with amino group-reactive (epoxy or acrylic) anchoring groups (Figure 1.29). Osada reported a combinatorial photo cross-linking of small molecules on affinity beads, where the small molecules were without a tether bound to the matrix at various sites (Kanoh et al., 2005).

On-chip synthesis provides a further improvement of efficiency, but has practical limitations in terms of flexibility and the quality of the *on the spot* synthesized compounds. Moreover, the homogeneity and the quality of these microarrays are difficult to control. Chemical libraries, applying compounds with similar functional groups and reactivity (like in the case of amino acids), were successfully used in making macroarrays on cellulose membranes using the *SPOT* synthesis concept (Frank, 2002).

1.6.3 Application of chemical microarrays in target identification

Several innovative techniques and applications were recently reported. Stockwell (Bailey, Sabatini, and Stockwell, 2004) developed microarrays of small molecules embedded in biodegradable polymers, while Kim and Park (2003) immobilized cGMP onto agarose beads. For parallel miniaturized enzyme assays, Gosalia and Diamond (2003) employed microarrays for liquid phase nanoliter reactions. Combinatorial libraries as individual nanodroplets were microarrayed onto glass slides and incubated with fluorogenic substrates (Gosalia et al., 2005), and the resulting fluorescent signal was scanned.

Schreiber gave another example where he successfully employed diversity-oriented synthesis for the discovery of transcription factor inhibitors (Koehler, Shamji, and Schreiber, 2003). Khersonsky and coworkers linked forward chemical genetics techniques with chemical microarrays. After phenotypic screening, triazine core-containing affinity matrices as microarrays were used (Khersonsky et al., 2003; Khersonsky and Chang, 2004).

1.6.3.1 Identification of disease-specific targets and small molecules with chemical microarrays: A reverse chemical genetics approach

In several diseases, specific genes (often oncogenes) are overexpressed, reflecting the specific genotype of the disease. Overexpression of genes or oncogenes and their protein products is often the best starting point for dissecting the pathways that lead to the development of specific diseases. Such proteins often serve as molecular targets particularly in cancer, where small molecules are designed to directly inhibit oncogenic proteins (oncoproteins) that are overexpressed. The direct inhibition of such gene product proteins (defined as oncoproteins in the case of cancer) and the relevant signaling pathways could provide novel opportunities for targeted therapy. Among the pioneering agents used in such targeted anticancer therapies were imatinib (Gleevec®) and trastuzumab (Herceptin®). Imatinib, as previously described, is a BCR-ABL tyrosine kinase inhibitor and useful in the treatment of CML and some solid tumors, whereas Herceptin is a monoclonal antibody developed against the Her2 oncoprotein, overexpressed in 25% of breast cancer cases (Goldenberg, 1999).

Identification and selection of disease-specific targets is one of the most important steps toward personalized medicine. The simultaneous identification of disease-specific protein targets and their small molecule binding partners, suitable as drug candidates after validation with a cell-based approach (a reverse chemical genetics approach), could radically reduce the timeline and costs of drug discovery and development. Diverse small molecule microarrays were applied to identify such proteins directly from cancer cell lysates/fractions in comparison with healthy counterparts as controls (Molnár et al., 2006). The chemical micro-arrays containing 600 small molecules were incubated with complete protein extracts of primary melanocytes and different melanoma cell lines (RVH and A375). Since both cell lines were universally labeled with fluorescent dyes, it allowed comparison of the binding intensity of the immobilized ligands. *Positive* small molecules were identified as they showed higher signals (more fluorescent-intensive interaction) in their position (spot) with proteins from melanoma cells than with proteins from primary melanocytes. Such positive small molecules (*fishing ligands*) were used for target isolation through affinity-based methods leading to the identification of tubullin as a primary binding protein and two other proteins previously not associated with cancer. In cell-based assays, one of the positive compounds showed significant cytotoxicity against various cancer cell lines (melanoma cells, HepG).

1.7 Summary

In the last 5 years, we have witnessed a rapid development in chemical genomics technologies. While the major principles remained the same, new subdisciplines emerged such as combination chemical genetics, drug chemical proteomics, and so forth. The major trends point toward the increasing involvement of system biology as well as network analysis, which helps to understand diseases. With more sophisticated and accurate analytical systems, multiple effects can be monitored as a consequence of the small molecules perturbation of the signaling network. Another important trend is that chemical genomics more frequently applies model organisms and advanced cell-based assays using engineered oncogenic cell lines to identify antitumor agents more rapidly. Although chemical genomics approaches are often linked to diagnostics, the implementation of personalized medicine is slower than expected. Regarding small molecules and chemistry, more and more laboratories utilize diverse, prefiltered, or annotated commercial libraries rather then spending much effort and time on tedious library synthesis.

References

Alaimo, P. J., Shogren-Knaak, M. A., and Shokat, K. M. 2001. Chemical genetic approaches for the elucidation of signaling pathways. *Curr. Opin. Chem. Biol.* 5:360–67.

Arya, P., Joseph, R., Gan, Z., and Rakic, B. 2005. Exploring new chemical space by stereocontrolled diversity-oriented synthesis. *Chem. Biol.* 12(2):163–80.

Bailey, S. N., Sabatini, D. M., and Stockwell, B. R. 2004. Microarrays of small molecules embedded in biodegradable polymers for use in mammalian cell-based screens. *Proc. Natl. Acad. Sci. USA* 101:16144–49.

Bantscheff, M., Eberhard, D., Abraham, Y., Bastuck, S., Boesche, M., Hobson, S., et al. 2007. Quantitative chemical proteomics reveals mechanisms of action of clinical ABL kinase inhibitors. *Nat. Biotechnol.* 25(9):1035–44.

Barglow, K. T. and Cravatt, B. F. 2007. Activity-based protein profiling for the functional annotation of enzymes. *Nat. Methods* 4(10):822–7.

Bon, R. S. and Waldmann, H. 2010. Bioactivity-guided navigation of chemical space. *Acc Chem Res.* 43(8):1103–14.

Browne, L. J., Furness, L. M., Natsoulis, G., Pearson, C., and Jarnagin, K. 2002. Chemogenomics: Pharmacology with genomics tools. *Targets* 1:59–65.

Burkard, M. E. and Jallepalli, P. V. 2010. Validating cancer drug targets through chemical genetics. *Biochim. Biophys. Acta* 1806(2):251–7.

Butcher, E. C., Berg, E. L., and Kunkel, E. J. 2004. Systems biology in drug discovery. *Nat. Biotechnol.* 22(10):1253–9.

Cai, S. X., Drewe, J., and Kasibhatla, S. 2006. A chemical genetics approach for the discovery of apoptosis inducers: From phenotypic cell-based HTs assay and structure-activity relationship studies, to identification of potential anticancer agents and molecular targets. *Curr. Med. Chem.* 13:2627–44.

Chen, H., Kovar, J., Sissons, S., Cox, K., Matter, W., Chadwell, F., et al. 2005. A cell-based immunocytochemical assay for monitoring kinase signaling pathways and drug efficacy. *Anal. Biochem.* 338:136–42.

Cordier, C., Morton, D., Murrison, S., Nelson, A., and O'Leary-Steele, C. 2008. Natural products as an inspiration in the diversity-oriented synthesis of bioactive compound libraries. *Nat. Prod. Rep.* 25(4):719–37.

Corsello, S. M., Roti, G., Ross, K. N., Chow, K. T., Galinsky, I., DeAngelo, D. J., et al. 2009. Identification of AML1-ETO modulators by chemical genomics. *Blood* 113(24):6193–205.

Cravatt, B. F., Wright, A. T., and Kozarich, J. W. 2008. Activity-based protein profiling: From enzyme chemistry to proteomic chemistry. *Ann. Rev. Biochem.* 77:383–414.

Croston, G. E. 2002. Functional cell-based uHTS in chemical genomic drug discovery. *Trends Biotechnol.* 20:110–15.

Csermely, P., Agoston, V., and Pongor, S. 2005. The efficiency of multi-target drugs: The network approach might help drug design. *Trends Pharm. Sci.* 26(4):178–82.

Curtis, T. K., Borisy, A. A., and Stockwell, B. R. 2005. Multicomponent therapeutics for networked systems. *Nature Rev. Drug Disc.* 4:71–78.

Darvas, F., Dormán, G., Ürge, L., Szabó, I., Rónai, Z., and Sasváry-Székely, M. 2001. Combinatorial chemistry: Facing the challenge of chemical genomics. *Pure Appl. Chem.* 73:1487–98.

Darvas, F., Keserű, G., Papp, Á., Dormán, G., Ürge, L., and Krajcsi, P. 2002. *In silico* and *ex silico* ADME approaches for drug discovery. *Curr. Top. Med. Chem.* 2:1269–77.

Darvas, F., Dormán, G., Puskás, L. G., Bucsai, A., and Ürge, L. 2004. Chemical genomics for fast and integrated target identification and lead optimization. *Med. Chem. Res.* 13:643–59.

Daub, H., Godl, K., Brehmer, D., Klebl, B., and Müller, G. 2004. Evaluation of kinase inhibitor selectivity by chemical proteomics. *Assay Drug Dev. Technol.* 2(2):215–24.

Dean, P. M. 2007. Chemical genomics: A challenge for *de novo* drug design. *Mol. Biotechnol.* 37(3):237–45.

Dickopf, S., Frank, M., Junker, H. D., Maier, S., Metz, G., Ottleben, H., et al. 2004. Custom chemical microarray production and affinity fingerprinting for the S1 pocket of factor VIIa. *Anal. Biochem.* 335:50–57.

Dolma, S., Lessnick, S. L., Hahn, W. C., and Stockwell, B. R. 2003. Identification of genotype-selective antitumor agents using synthetic lethal chemical screening in engineered human tumor cells. *Cancer Cell* 3(3):285–96.

Dormán, G. and Darvas, F. 2004. Utilizing Small Molecules in Chemical Genomics: Toward HT Approaches. In *Chemical Genomics*, ed. F. Darvas, A. Guttman, and G. Dormán, 137–197. New York, Basel: Marcel Dekker.

Drahl, C., Cravatt, B. F., and Sorensen, E. J. 2005. Protein-reactive natural products. *Angew. Chem. Int. Ed. Engl.* 44(36):5788–809.

Duffner, J. L., Clemons, P. A., and Koehler, A. N. 2007. A pipeline for ligand discovery using small-molecule microarrays. *Curr. Opin. Chem. Biol.* 11:74–82.

Erlanson, D. A. 2006. Fragment-based lead discovery: A chemical update. *Curr. Opin. Biotechnol.* 17(6):643–52.

Evans, M. J. and Cravatt, B. F. 2006. Mechanism-based profiling of enzyme families. *Chem. Rev.* 106(8):3279–301.

Fabian, M. A., Biggs, W. H. 3rd, Treiber, D. K., Atteridge, C. E., Azimioara, M. D., Benedetti, M. G., et al. 2005. A small molecule-kinase interaction map for clinical kinase inhibitors. *Nat. Biotechnol.* 23(3):329–36.

Fedorov, O., Sundström, M., Marsden, B., and Knapp, S. 2007. Insights for the development of specific kinase inhibitors by targeted structural genomics. *Drug Discov. Today* 12(9–10):365–72.

Frank, R. 2002. The SPOT-synthesis technique: Synthetic peptide arrays on membrane supports—Principles and applications. *J. Immunol. Methods* 267:13–26.

Freundlieb, S. and Gamer, J. 2002. Chemical microarrays: A novel approach to drug discovery. *New Drugs* 3:54–60.

Furka, Á., Sebestyén, F., Asgedom, M., and Dibó, G. 1991. General method for rapid synthesis of multicomponent peptide mixtures. *Int. J. Pept. Prot. Res.* 37:487–93.

Ganter, B., Snyder, R. D., Halbert, D. N., and Lee, M. D. 2006. Toxicogenomics in drug discovery and development: Mechanistic analysis of compound/class-dependent effects using the DrugMatrix database. *Pharmacogenomics* 7(7):1025–44.

Goldenberg, M. M. 1999. Trastuzumab, a recombinant DNA-derived humanized monoclonal antibody, a novel agent for the treatment of metastatic breast cancer. *Clin. Ther.* 21:309–18.

Gonzalez-Angulo, A. M., Hennessy, B. T., and Mills, G. B. 2010. Future of personalized medicine in oncology: A systems biology approach. *J. Clin. Oncol.* 28(16):2777–83.

Gosalia, D. N. and Diamond, S. L., 2003. Printing chemical libraries on microarrays for fluid phase nanoliter reactions. *Proc. Natl. Acad. Sci. USA* 100:8721–26.

Gosalia, D. N., Salisbury, C. M., Ellman, J. A., and Diamond, S. L. 2005. High throughput substrate specificity profiling of serine and cysteine proteases using solution-phase fluorogenic peptide microarrays. *Mol. Cell. Proteomics* 4:626–36.

Gray, N. S., Wodicka, L., Thunnissen, A. W. H., Norman, T. C., Kwon, S., Espinoza, H. F., et al. 1998. Exploiting chemical libraries, structure, and genomics in the search for kinase inhibitors. *Science* 281:533–35.

Guo, W., Twomey, D., Lamb, J., Schlis, K., Agarwal, J., Stam, R. W., et al. 2006. Gene expression-based chemical genomics identifies rapamycin as a modulator of MCL1 and glucocorticoid resistance. *Cancer Cell* 10(4):331–42.

Hackler, L., Jr., Dormán, G., Kele, Z., Ürge, L., Darvas, F., and Puskás, L. G. 2003. Development of chemically modified glass surfaces for nucleic acid, protein and small molecule microarrays. *Mol. Div.* 7:25–33.

Hagenstein, M. C. and Sewald, N. 2006. Chemical tools for activity-based proteomics. *J. Biotechnol.* 124:56–73.

Hantschel, O., Rix, U., and Superti-Furga, G. 2008. Target spectrum of the BCR-ABL inhibitors imatinib, nilotinib, and dasatinib. *Leuk. Lymph.* 49(4):615–19.

Hatanaka, Y. and Sadakane, Y. 2002. Photoaffinity labeling in drug discovery and developments: Chemical gateway for entering proteomic frontier. *Curr. Top. Med. Chem.* 2:271–88.

Irwin, J. J. and Shoichet, B. K. 2005. ZINC—A free database of commercially available compounds for virtual screening. *J. Chem. Inf. Model.* 45(1):177–82.

Joseph, R. and Arya, P. 2004. Combinatorial Chemistry in the Age of Chemical Genomics. In *Chemogenomics in Drug Discovery*, ed. H. Kubinyi and G. Müller, 405–433. Weinheim: Wiley-VCH.

Kanoh, N., Honda, K., Simizu, S., Muroi, M., and Osada, H. 2005. Photo-cross-linked small-molecule affinity matrix for facilitating forward and reverse chemical genetics. *Angew. Chem. Int. Ed. Engl.* 44(23):3559–62.

Kapoor, T. M., Mayer, T. U., Coughlin, M. L., and Mitchison, T. J. 2000. Probing spindle assembly mechanisms with monastrol, a small molecule inhibitor of the mitotic kinesin, Eg5. *J. Cell. Biol.* 150:975–88.

Kau, T. R., Schroeder, F., Ramaswamy, S., Wojciechowski, C. L., Zhao, J. J., Roberts, T. M., et al. 2003. A chemical genetic screen identifies inhibitors of regulated nuclear export of a Forkhead transcription factor in PTEN-deficient tumor cells. *Cancer Cell* 4(6):463–76.

Kessmann, H. and Ottleben, H. A. 2001. A shortcut on the long road to drug discovery. *Innov. Pharm. Technol.* 24–27.

Khersonsky, S. M. and Chang, Y. T. 2004. Forward chemical genetics: Library scaffold design. *Comb. Chem. High-Throughput Screen.* 7:645–52.

Khersonsky, S. M., Jung, D. W., Kang, T. W., Walsh, D. P., Moon, H. S., Jo, H., et al. 2003. Facilitated forward chemical genetics using tagged triazine library and zebrafish embryo screening. *J. Am. Chem. Soc.* 125:11804–805.

Khersonsky, S. M. and Chang, Y. T. 2004. Strategies for facilitated forward chemical genetics. *ChemBioChem.* 5:903–08.

Khor, T. O., Ibrahim, S., and Kong, A. N. 2006. Toxicogenomics in drug discovery and drug development: Potential applications and future challenges. *Pharm. Res.* 23(8):1659–64.

Kim, E. and Park, J. M. 2003. Identification of novel target proteins of cyclic GMP signaling pathways using chemical proteomics. *J. Biochem. Mol. Biol.* 36:299–304.

Koehler, A. N., Shamji, A. F., and Schreiber, S. L. 2003. Discovery of an inhibitor of a transcription factor using small molecule microarrays and diversity-oriented synthesis. *J. Am. Chem. Soc.* 125:8420–21.

Kruse, U., Bantscheff, M., Drewes, G., and Hopf, C. 2008. Chemical and pathway proteomics: Powerful tools for oncology drug discovery and personalized health care. *Mol. Cell. Proteomics* 7(10):1887–901.

Kuruvilla, F. G., Shamji, A. F., Sternson, S. M., Hergenrother, P. J., and Schreiber, S. L. 2002. Dissecting glucose signaling using diversity-oriented synthesis and small-molecule microarrays. *Nature* 416(6881):653–57.

Kwon, H. J. 2006. Discovery of new small molecules and targets toward angiogenesis via chemical genomics approach. *Curr. Drug Targets* 7:397–405.

Kwon, H. J., Owa, T., Hassig, C. A., Shimada, J., and Schreiber, S. L. 1998. Depudecin induces morphological reversion of transformed fibroblasts via the inhibition of histone deacetylase. *Proc. Natl. Acad. Sci. USA* 95:3356–61.

Kwon, O., Park, S., B., and Schreiber, S. L. 2002. Skeletal diversity via a branched pathway: Efficient synthesis of 29 400 discrete, polycyclic compounds and their arraying into stock solutions. *J. Am. Chem. Soc.* 124(45):13402–4.

Lampson, M. A. and Kapoor, T. M. 2006. Unraveling cell division mechanisms with small-molecule inhibitors. *Nat. Chem. Biol.* 2(1):19–27.

Lehár, J., Zimmermann, G. R., Krueger, A. S., Molnar, R. A., Ledell, J. T., Heilbut, A. M., et al. 2007. Chemical combination effects predict connectivity in biological systems. *Mol. Syst. Biol.* 3:80. doi:10.1038/msb4100116.

Lehár, J., Stockwell, B. R., Giaever, G., and Nislow, C. 2008. Combination chemical genetics. *Nat. Chem. Biol.* 4(11):674–81.

Lipinski, C. A., Lombardo, F., Dominy, B. W., and Feeney, P. J. 1997. Experimental and computational approaches to estimate solubility and permeability in drug discovery and development settings. *Adv. Drug Deliv. Rev.* 23:3–25.

MacBeath, G., Koehler, A. N., and Schreiber, S. L. 1999. Printing small molecules as microarrays and detecting protein-ligand interactions en masse. *J. Am. Chem. Soc.* 121:7967–68.

MacBeath, G. and Schreiber, S. L. 2000. Printing proteins as microarrays for high-throughput function determination. *Science* 289(5485):1760–3.

MacBeath, G. 2001. Chemical genomics: What will it take and who gets to play? *Genome Biology* 2: Comment 2005.1. doi:10.1186/gb-2001-2-6-comment2005.

MacRae, C. A. and Peterson, R. T. 2003. Zebra fish-based small molecule discovery. *Chem. Biol.* 10(10):901–8.

Marcaurelle, L. A. and Johannes, C. W. 2008. Application of natural product-inspired diversity-oriented synthesis to drug discovery. *Prog. Drug Res.* 66:187, 189–216.

Mayer, T. U., Kapoor, T. M., Haggarty, S. J., King, R. W., Schreiber, S. L., and Mitchison, T. J. 1999. Small molecule inhibitor of mitotic spindle bipolarity identified in a phenotype-based screen. *Science* 286:971–74.

Mitsopoulos, G., Walsh, D. P., and Chang, Y. T. 2004. Tagged library approach to chemical genomics and proteomics. *Curr. Opin. Chem. Biol.* 8:26–32.

Molnár, E., Hackler, L., Jr., Jankovics, T., Ürge, L., Darvas, F., and Fehér, L. Z. 2006. Application of small molecule microarrays in comparative chemical proteomics. *QSAR Comb. Sci.* 25(11):1020–27.

Oprea, T. and Gottfries, J. 1999. Toward minimalistic modeling of oral drug absorption. *J. Mol. Graph. Model.* 17:261–74.

Pal, K. 2000. The keys to chemical genomics. *Mod. Drug Discov.* 3:47–55.

Paulick, M. G. and Bogyo, M. 2008. Application of activity-based probes to the study of enzymes involved in cancer progression. *Curr. Opin. Genet. Dev.* 18(1):97–106.

Peták, I., Schwab, R., Orfi, L., Kopper, L., and Kéri, G. 2010. Integrating molecular diagnostics into anticancer drug discovery. *Nat. Rev. Drug Discov.* 9(7):523–35.

Petrelli, A. and Giordano, S. 2008. From single- to multi-target drugs in cancer therapy: When aspecificity becomes an advantage. *Curr. Med. Chem.* 15:422–32.

Potuzak, J. S., Moilanen, S. B., and Tan, D. S. 2004. Discovery and applications of small molecule probes for studying biological processes. *Gen. Eng. Rev.* 21:11–55.

Reayi, A. and Arya, P. 2005. Natural product-like chemical space: Search for chemical dissectors of macromolecular interactions. *Curr. Opin. Chem. Biol.* 9(3):240–7.

Rix, U. and Superti-Furga, G. 2009. Target profiling of small molecules by chemical proteomics. *Nat. Chem Biol.* 5(9):616–24.

Root, D. E., Flaherty, S. P., Kelley, B. P., and Stockwell, B. R. 2003. Biological mechanism profiling using an annotated compound library. *Chem. Biol.* 10(9):881–92.

Sachinidis, A., Sotiriadou, I., Seelig, B., Berkessel, A., and Hescheler, J. 2008. A chemical genetics approach for specific differentiation of stem cells to somatic cells: A new promising therapeutical approach. *Comb. Chem. High Throughput Screen.* 11:70–82.

Schreiber, S. L. 1998. Chemical genetics resulting from a passion for synthetic organic chemistry. *Bioorg. Med. Chem.* 6:1127–52.

Schuffenhauer, A., Ertl, P., Roggo, S., Wetzel, S., Koch, M. A., and Waldmann, H. 2007. The scaffold tree—Visualization of the scaffold universe by hierarchical scaffold classification. *J. Chem. Inf. Model.* 47(1):47–58.

Shawn, H., St. Onge, R. P., Giaever, G., and Nislow, C. 2008. Yeast chemical genomics and drug discovery: An update. *Trends Pharm. Sci.* 29(10):499–504.

Shingo, D., Tsunoda, T., Kitahara, O., Yanagawa, R., Zembutsu, H., Katagiri, T., et al. 2002. An integrated database of chemosensitivity to 55 anticancer drugs and gene expression profiles of 39 human cancer cell lines. *Cancer Res.* 62:1139–47.

Stockwell, B. R. 2000. Frontiers in chemical genetics. *Trends Biotechnol.* 18: 449–55.

Stockwell, B. R., Haggarty, S. J., and Schreiber, S. L. 1999. High-throughput screening of small molecules in miniaturized mammalian cell-based assays involving post-translational modifications. *Chem. Biol.* 6(2):71–83.

Stockwell, B. R., Hardwick, J. S., Tong, J. K., and Schreiber, S. L. 1999. Chemical genetic and genomic approaches reveal a role for copper in specific gene activation. *J. Am. Chem. Soc.* 121:10662–63.

Tan, S. 2005. Diversity-oriented synthesis: Exploring the intersections between chemistry and biology. *Nat. Chem. Biol.* 1:74–84.

Tan, D., Foley, M. A., Stockwell, B. R., Shair, M. D., and Schreiber, S. L. 1999. Synthesis and preliminary evaluation of a library of polycyclic small molecules for use in chemical genetic assays. *J. Am. Chem. Soc.* 121:9073–87.

Terstappen, G. C., Schlüpen, C., Raggiaschi, R., and Gaviraghi, G. 2007. Target deconvolution strategies in drug discovery. *Nat. Rev. Drug Discov.* 6:891–903.

Umarye, J. D., Lessmann, T., García, A. B., Mamane, V., Sommer, S., and Waldmann, H. 2007. Biology-oriented synthesis of stereochemically diverse natural-product-derived compound collections by iterative allylations on a solid support. *Chemistry* 13(12):3305–19.

Uttamchandani, M., Wang, J., and Yao, S. Q. 2006. Protein and small molecule microarrays: Powerful tools for high-throughput proteomics. *Mol. Biol. Syst.* 2:58–68.

Vegas, A. J., Fuller, J. H., and Koehler, A. N. 2008. Small-molecule microarrays as tools in ligand discovery. *Chem. Soc. Rev.* 37(7):1385–94.

Waters, M., Stasiewicz, S., and Merrick, B. A. 2008. CEBS—Chemical effects in biological systems: A public data repository integrating study design and toxicity data with microarray and proteomics data. *Nucleic Acids Res.* 36:D892–D900.

Witucki, L. A., Huang, X., Shah, K., Liu, L., Kyin, S., Eck, M. J., and Shokat, K. M. 2002. Mutant tyrosine kinases with unnatural nucleotide specificity retain the structure and phospho-acceptor specificity of the wild-type enzyme. *Chem. Biol.* 9:25–33.

Yagoda, N., von Rechenberg, M., Zaganjor, E., Bauer, A. J., Yang, W. S., Fridman, D. J., et al. 2007. RAS-RAF-MEK-dependent oxidative cell death involving voltage-dependent anion channels. *Nature* 447(7146):864–8.

Yang, W. S. and Stockwell, B. R. 2008. Synthetic lethal screening identifies compounds activating iron-dependent, nonapoptotic cell death in oncogenic-rasharboring cancer cells. *Chem. Biol.* 15(3): 234–45.

Yildirim, M. A., Goh, K. I., Cusick, M. E., Barabási, A. L., and Vidal, M. 2007. Drug-target network. *Nat. Biotechnol.* 22(10):1119–26.

Zheng, X. S. F. and Chan, T. 2002. Chemical genomics in the global study of protein functions. *Drug Discov. Today* 7:197–205.

chapter two

Development and application of novel analytical methods in lipidomics

Yuqin Wang, Anna Meljon, Michael Ogundare, and William J. Griffiths

Contents

2.1 Lipidomics in the postgenomic era

Following the sequencing of the human, and many other genomes, scientific interest shifted toward analysis of the proteome and metabolome. Major efforts in metabolomics have now begun. The primary analytical techniques used in these endeavors are nuclear magnetic resonance (NMR) spectrometry and mass spectrometry (MS) [1,2]. Both techniques offer their own advantages and disadvantages, but in this chapter we will confine our discussion to MS methods. Mass spectrometry has been used for metabolite analysis over several decades, initially in the guise of direct

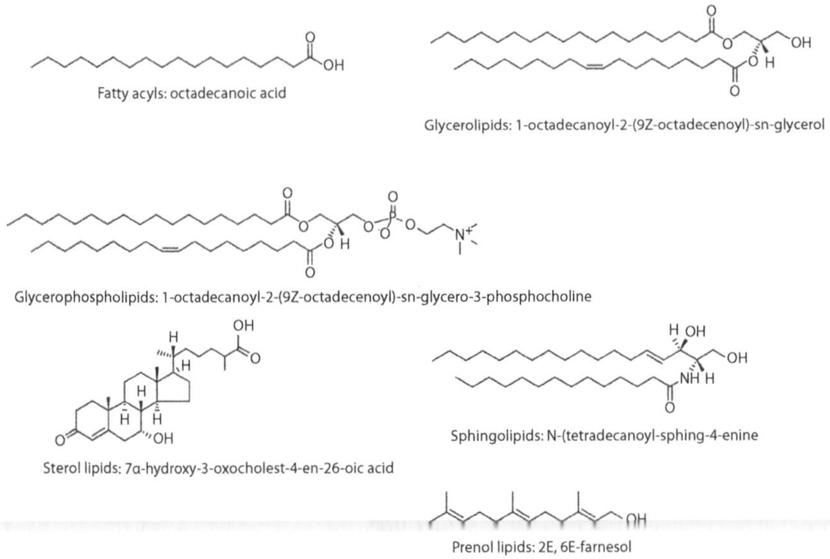

Fatty acyls: octadecanoic acid

Glycerolipids: 1-octadecanoyl-2-(9Z-octadecenoyl)-sn-glycerol

Glycerophospholipids: 1-octadecanoyl-2-(9Z-octadecenoyl)-sn-glycero-3-phosphocholine

Sterol lipids: 7α-hydroxy-3-oxocholest-4-en-26-oic acid

Sphingolipids: N-(tetradecanoyl-sphing-4-enine

Prenol lipids: 2E, 6E-farnesol

Figure 2.1 Structures of the six classes of lipids, usually analyzed in mass spectrometry-based lipidomics.

inlet electron ionization (EI)-MS or gas chromatography (GC)-MS [3,4], and more recently as direct infusion-electrospray ionization (DI-ESI)-MS [5,6] or liquid chromatography (LC)-ESI-MS [7–9]. The requirement for volatility and thermal stability when EI-MS and GC-MS are used dictates that many biological analytes require derivatization before analysis [10], while ESI-MS and LC-ESI-MS both require the analyte to be soluble in the spraying solvent [11]. These requirements are similarly true for the analogous ionization methods of chemical ionization (CI) and atmospheric pressure CI (APCI), respectively. The respective needs of volatility and solubility argue against a metabolome-wide analysis using a single experimental platform on account of the extreme diversity of chemical and physical properties of different metabolites. As a consequence of their solubility in organic solvents, lipids constitute one category of metabolites that require separate analysis from water-soluble analytes (Figure 2.1), hence the emergence of a separate *omic* division, namely, lipidomics [12–14].

Fundamentally, lipidomics can be defined as the qualitative and quantitative determination of all lipid molecular species in a biological fluid, tissue, or cell type. This goal is impractical for individual groups of workers, but is a realistic aim, at least for the more abundant lipids, for consortia of lipid scientists. Such consortia exist and the Lipid Maps (http://www.lipidmaps.org/) grouping in the United States has already published results from extensive joint studies of the human plasma lipidome

[15] and a mouse macrophage cell line [16,17]. A more conservative goal in lipidomics is to compare batches of samples and to define differences in their lipidome following treatment or according to phenotype. Such analysis can be used in an attempt to identify biomarkers of disease state, disease progression, or response to therapy [18,19].

2.2   Global, shotgun, and top-down lipidomics

Shotgun lipidomics is a global lipidomic methodology pioneered by Han and Gross at Washington University [19–23] and by Shevchenko and colleagues in Dresden [18,24]. The essential feature of shotgun lipidomics is that it involves direct-infusion ESI-MS and tandem mass spectrometry (MS/MS) following a simple extraction step, in the absence of online chromatographic separation. Han and Gross call their version of shotgun lipidomics *multidimensional mass spectrometry* (MDMS). They combine the use of direct-infusion ESI-MS in positive- and negative-ion modes with MS/MS (see Figure 2.2), and in their current studies are able to screen for 36,000 potential lipid molecular species [23]. Shotgun lipidomics, like all MS-based analysis, is biased toward compounds that are most abundant and easily ionized. As a result, shotgun spectra tend to be dominated by glycerolipids, glycerophospholipids, and sphingolipids (see Figure 2.1).

A generic MDMS shotgun lipidomic experiment is based on the following steps: (i) Lipid extraction into an organic solvent, typically by a modified Bligh and Dyer method [25]; (ii) Direct-infusion ESI-MS without and with the addition of LiOH in the negative- and positive-ESI modes; (iii) Acquisition of head-group and acyl chain-specific neutral-loss and precursor-ion scans when using tandem quadrupole instruments, or product-ion scans when using Q-TOF instruments (Figure 2.2); and (iv) Data analysis including peak annotation and quantification. MDMS was originally pioneered on tandem quadrupole instruments that excel in the acquisition of neutral-loss and precursor-ion scans. These scans fit well with the analysis of glycerophospholipids, glycerolipids, and sphingolipids, where the different classes of lipids give characteristic head-group fragments, while fatty acyl groups are characterized by their RCO_2^- ions (Table 2.1, Figure 2.3). In an MDMS analysis the first (x) dimension is a mass (m/z) scan, also called a *survey scan*, while successive neutral-loss and precursor-ion scans represent the second (y) dimension. This is illustrated in Figure 2.4, which shows MDMS analysis of a lipid extract of mouse retina [22].

The advantage of MDMS, particularly when using tandem quadrupole instruments, is that MS/MS analysis time is only utilized on fragment-ions or neutral-losses of interest, avoiding the generation of redundant data with the inherent time and sensitivity penalties. The regular ordering of lipids in nature with well-defined head groups and a limited number of

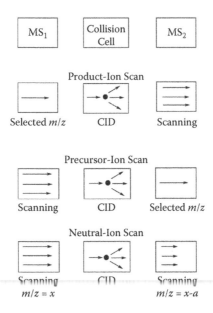

Figure 2.2 Scan types utilized in lipidomic analysis by ESI-MS/MS. An MS/MS instrument consists of an initial mass (m/z) analyzer (MS_1), a collision cell, and a second mass (m/z) analyzer (MS_2). The two mass (m/z) analyzers and collision cell are separated *in space* on a beam instrument, such as tandem quadrupoles and Q-TOFs, and *in time* in ion traps. Product-ion, precursor-ion, and neutral-loss scans are performed by respectively scanning MS_2, MS_1, or MS_1 and MS_2 in parallel. Multiple reaction monitoring (MRM) chromatograms are recorded with MS_1 and MS_2 fixed for transitions of interest. MS^3 or MS/MS/MS spectra are recorded when a third mass (m/z) analyzer MS_3 is utilized following a second collision cell. MS^3 and further MS^n spectra are often recorded on ion-trap instruments.

acyl chains make this possible (Table 2.1, Figure 2.3). Shotgun lipidomics can also be performed using Q-TOF-type instruments [26,27], in which case every mass (m/z) in the survey scan is subjected to MS/MS. Q-TOF instruments offer sensitivity advantages for the acquisition of product-ion spectra in that all fragment-ions are detected *at once* as opposed to *one at a time* on quadrupole instruments.

Shevchenko and colleagues have developed a somewhat different form of shotgun lipidomics which is particularly suitable to applications utilizing a hybrid linear ion trap (LIT) Fourier transform (FT) instrument, for example, LTQ-Orbitrap, called *top-down* lipidomics [18,24]. In a top-down lipidomics experiment, the survey scan is recorded at high resolution (100,000) (full width at half maximum height [FWMH]) in the FT analyzer. Mass accuracy on FT instruments is typically <5 ppm and often with internal calibration <2 ppm. This degree of mass accuracy

Table 2.1 MS/MS Scans for Phospholipid Analysis

Class[1]	Precursor-Ion	Scan[2]	Fragment[3]	Structure
CL	$[M-2H]^{2-}$	PS	m/z 153	$[HO_3POCH_2CH(O)CH_2]^-$
GPA	$[M-H]^-$	PS	m/z 153	$[HO_3POCH_2CH(O)CH_2]^-$
GPCho	$[M+H]^+$	PS	m/z 184	$[H_2O_3POCH_2CH_2N(CH_3)_3]^+$
	$[M+Cl]^-$	PS	m/z 50	(CH_3Cl)
GPEtn	$[M-H]^-$	PS	m/z 196	$[CH_2(O)CHCH_2OPO_3C_2H_4NH_2]^-$
GPGro	$[M-H]^-$	PS	m/z 153	$[HO_3POCH_2CH(O)CH_2]^-$
		PS	m/z 227	$[CH_2(O)CHCH_2OPO_3CH_2CH(OH)CH_2OH]^-$
GPIns	$[M-H]^-$	PS	m/z 153	$[HO_3POCH_2CH(O)CH_2]^-$
		PS	m/z 241	$[PO_4C_6H_{10}O_4]^-$
GPSer	$[M-H]^-$	PS	m/z 153	$[HO_3POCH_2CH(O)CH_2]^-$
		NL	87 Da	$(CH_2C(NH_2)CO_2H)$
SM	$[M+H]^+$	PS	m/z 184	$[H_2O_3POCH_2CH_2N(CH_3)_3]^+$
	$[M+Cl]^-$	PS	m/z 50	(CH_3Cl)

[1] Fahy, E. et al., 2005, A Comprehensive Classification System for Lipids, *J. Lipid Res.* 46:839–61.
[2] Precursor-ion scan (PS); Neutral-loss scan (NL).
[3] For more extensive tables/information see Han and Gross [21], Griffiths [29], and Pulfer and Murphy [30].

gives information on elemental composition [31]. Although an elemental composition will not exactly define a lipid, it will greatly reduce the number of possible structures. Further characterization of lipids of interest can be achieved by utilizing MS/MS (MS²) or MSⁿ. In a recent report, Graessler et al. presented results from the shotgun top-down analysis of blood plasma where they compared the lipid profile of plasma hypertensive patients and controls [18]. They were able to identify 95 lipid species, which were predominantly glycerophosphocholines (GPCho), glycerophosphoethanolamines (GPEtn), phosphosphingolipids (SM), di- and tri-radyl glycerols (DG, TG), and cholesterol esters (Figure 2.5). In a similar study on HeLa cell extracts, Schwudke et al. [24] were able to assign chemical compositions to 262 lipid peaks.

Quantification in shotgun lipidomics requires the addition of internal standards. Internal standards should be added during the extraction step and should consist of at least one standard for each lipid class.

Shotgun lipidomics has been predominantly used with a minimum of sample preparation. However, by utilizing offline chromatographic separation, the coverage of the lipidome can be greatly extended. We have exploited reversed phase (RP) solid phase extraction (SPE) to target conjugated steroids, sterols, and bile alcohols/acids in plasma, which we have then analyzed by shotgun top-down *steroidomics*.

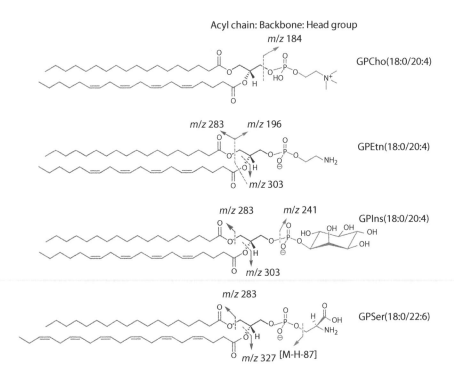

Figure 2.3 MS/MS fragmentation pattern of some glycerophospholipids (see also Table 2.1). Nomenclature and abbreviations are those recommended by Fahy. (Fahy, E. et al., 2005, A Comprehensive Classification System for Lipids, *J. Lipid Res.* 46:839–61.)

2.3   *Shotgun steroidomics*

Bile acids (ng/mL–μg/mL), bile alcohols (ng/mL) (also called *oxysterols*), and hormonal steroids (ng/mL–μg/mL) are present in plasma in their free and conjugated forms at significant levels [32,33]. Their levels may vary as a result of disease, for example, liver dysfunction, or as a consequence of an inborn error of metabolism. The idea of performing a shotgun analysis of the plasma steroidome is not new. Shackleton and colleagues performed such studies on plasma and urine in the early 1980s following the introduction of fast atom bombardment (FAB)-MS [33–36]. This was made possible at the time by their use of the newly launched C_{18} SPE cartridge that allowed the preparation of a steroid-rich fraction for subsequent FAB-MS analysis. That and the fact that steroid conjugates of sulphuric acid are readily ionized by negative-ion FAB made plasma analysis of such steroids such as dehydroepiandrosterone (DHEA)-sulphate comparatively straightforward [36]. Despite this, steroids are poorly represented in modern metabolomic [9,37] and lipidomic [15,18] investigations of plasma.

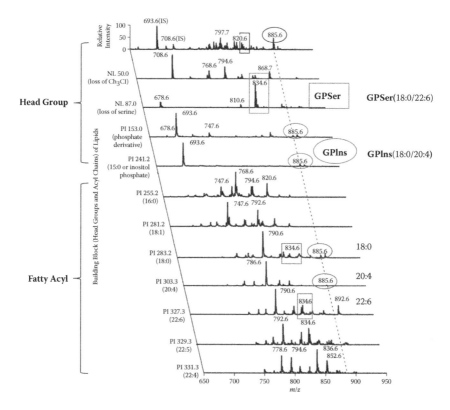

Figure 2.4 MDMS of a chloroform extract of mouse retina prepared by a modified Bligh and Dyer method [25]. Direct-infusion ESI spectra were acquired directly from the diluted lipid extract (total lipid ~ 50 pmol/μL) using a TSQ Quantum Ultra tandem quadrupole instrument. The first (x) dimension is represented in the top panel (negative-ion ESI-MS). The second dimension (y) is represented by successive neutral-loss (NL) and precursor-ion (PI) scans. All traces were displayed after normalization to the base peak in each spectrum. Internal standard (IS); m:n, acyl chain containing m carbons and n double bonds. (Modified with permission from Han, X. et al., 2005, Shotgun Lipidomics of Phosphoethanolamine-Containing Lipids in Biological Samples after One-Step *in Situ* Derivatization, *J. Lipid Res.* 46:1548–60.)

However, the utilization of C_{18} SPE prior to shotgun negative-ESI-MS, or online C_{18} LC-ESI-MS, allows the facile analysis of steroid sulphates [38]. Shown in Figure 2.6 is a negative-ion ESI-MS of the steroid fraction from infant plasma (Figure 2.7).

Besides hormonal steroid conjugates (Figure 2.7), bile alcohol and bile acid conjugates can also be analyzed by shotgun steroidomics. These metabolites are often elevated in abundance in plasma as a consequence of an inborn error of bile acid biosynthesis or as a consequence

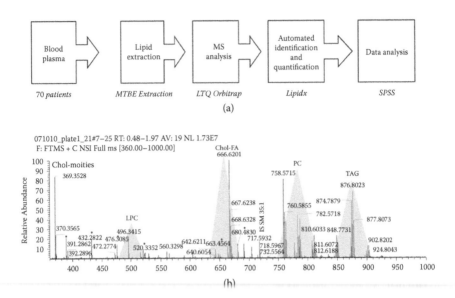

(a)

(b)

Figure 2.5 Top-down lipidomics strategy as exemplified by the analysis of human plasma. Plasma was extracted into methyl-*tert*-butyl ether. Extracts were redissolved in $CHCl_3$:CH_3OH:$CH_3CH(OH)CH_3$ 1:2:4 (v:v:v) containing 7.5 mM $NH_4CH_3CO_2$ and directly infused by ESI into an LTQ Orbitrap (Thermo Fisher). Spectra were recorded at high resolution (100,000 FWHM) and peaks annotated following exact mass measurement. Found in the shaded areas are lipids corresponding to the designated class: Lysophosphatidylcholine (LPC), sphingomyelin (SM), phosphatidylcholine (PC), triacylglycerol (TAG). (Reproduced with permission from Graessler J. et al., 2009, Top-Down Lipidomics Reveals Ether Lipid Deficiency in Blood Plasma of Hypertensive Patients, *PLoS One* 4:e62612009, July 15:4(7):e6261.)

of liver disease [39]. Inborn errors of bile acid biosynthesis, where an enzyme in the metabolic pathway is defective, lead to increased abundance of the metabolite preceding the blockage (Figures 2.8 and 2.9, Table 2.2), often allowing a fast and simple diagnosis. This, however, is not always the case as a blockage may result in a buildup of not only the directly preceding metabolite but also its precursors, resulting in a complicated spectrum.

2.4 Matrix-assisted laser desorption/ionization, desorption ionization on silicone, and desorption electrospray ionization

Although most lipidomic studies are conducted using ESI, other ionizations methods can also be used. Schiller et al. have recently reviewed the

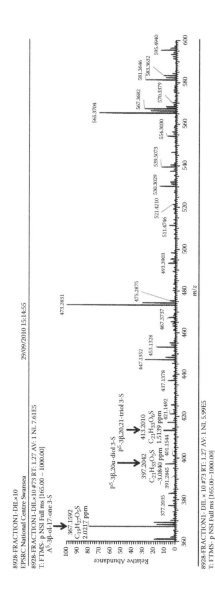

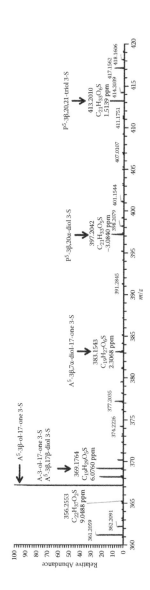

Figure 2.6 Shotgun steroidomic analysis of infant plasma. Steroids were isolated from infant plasma by SPE on a C$_{18}$ cartridge. Direct negative-ion ESI-MS analysis was performed on an LTQ Orbitrap XL (Thermo Fisher). Peaks which correspond to steroid sulphates are indicated. Steroid structures (see Figure 2.7) are postulated but have not been confirmed. (A) Androstane:; (p) Pregnane, location of double bonds indicated by a superscript.

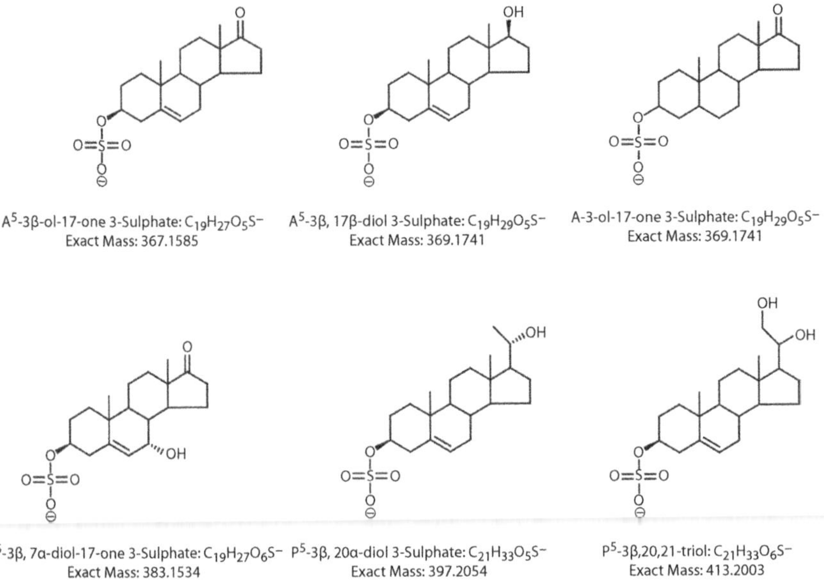

Chemical genomics and proteomics

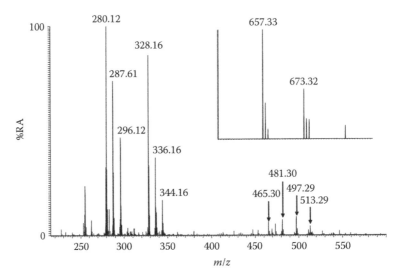

A5-3β-ol-17-one 3-Sulphate: C₁₉H₂₇O₅S⁻
Exact Mass: 367.1585

A5-3β, 17β-diol 3-Sulphate: C₁₉H₂₉O₅S⁻
Exact Mass: 369.1741

A-3-ol-17-one 3-Sulphate: C₁₉H₂₉O₅S⁻
Exact Mass: 369.1741

A5-3β, 7α-diol-17-one 3-Sulphate: C₁₉H₂₇O₆S⁻
Exact Mass: 383.1534

P5-3β, 20α-diol 3-Sulphate: C₂₁H₃₃O₅S⁻
Exact Mass: 397.2054

P5-3β,20,21-triol: C₂₁H₃₃O₆S⁻
Exact Mass: 413.2003

Figure 2.7 Structures postulated for the steroid sulphates. Annotated and abbreviations as described in Figure 2.6.

Figure 2.8 Shotgun negative-ion ESI spectrum of C₁₈ extracts of plasma from an infant with neonatal hepatitis of unknown etiology. The inset shows the extended *m/z* range 640–700. Spectra were recorded on an LTQ Orbitrap XL. The postulated identity of the peaks is given below and in Figure 2.9. %RA, % relative abundance.

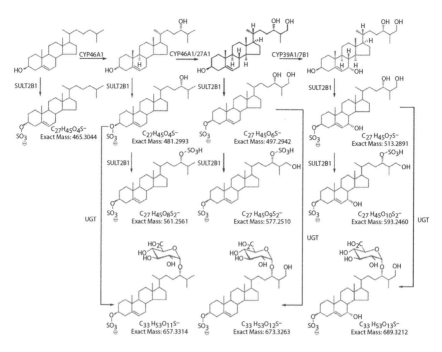

Figure 2.9 Suggested structures of the sterol peaks in Figure 2.8. The top row of the scheme shows necessary metabolic intermediates. For simplicity the conjugated sterols are drawn as the [M-H]⁻ ions. The location of the free hydroxyl and glucuronide groups have not been determined and are shown in the scheme by way of illustration of possible metabolic routes. Even by MS/MS, it is unlikely that the location of substituents could be determined. To achieve this it would be necessary to resort to keV collision-induced dissociation. See Griffiths [29], Sjövall, Griffiths, Setchell, et al. [32] and Makin, Honour, Shackleton, et al. [33]. Stereochemistry can be determined by GC-MS following solvolysis, hydrolysis, and derivatisation. See Sjövall, Griffiths, Setchell, et al. [32], and Makin, Honour, Shackleton, et al. [33].

application of matrix-assisted laser desorption/ionization (MALDI) in lipidomics [40], while Siuzdak and Cooks' groups have used desorption ionization on silicone (DIOS) [37], and desorption electrospray ionization (DESI), respectively, for metabolite, including lipid analysis [41]. DESI and MALDI offer the attractive feature of allowing *in situ* analysis, where ionization can be performed directly from a biological sample or tissue. However, they suffer from the general MS bias, which is toward the most easily ionized and abundant molecules and against those less easy to ionize and of lower abundance.

Table 2.2 Annotation of Cholesterol Metabolites in Figure 2.8

Peak *m/z*	Structure	Comment
465.30	Cholesterol sulphate	Conjugates, probably
481.30	Cholestenediol sulphate	of 3β(mono-tetra)
497.29	Cholestenetriol sulphate	hydroxy-5-ene
513.29	Cholestenetetrol sulphate	3β-sulphate
561.26/280.12	Cholestenediol disulphate	
576.23/287.61	Dihydroxycholenoyl taurine sulphate	
577.25/288.12	Cholestenetriol disulphate	
593.25/296.12	Cholestenetetrol disulphate	
657.33/328.16	Cholestenediol sulphate, glucuronide	
673.33/336.16	Cholestenetriol sulphate, glucuronide	
689.32/344.16	Cholestenetetrol sulphate, glucuronide	

2.5 Gas chromatography mass spectrometry (GC-MS)

Prior to the introduction of ESI, GC-MS was the most common MS method for lipid analysis. It is still extensively used today in metabolomic studies [42,43] and in studies of specific classes of lipids, essentially in a targeted approach [32,33] (see below).

2.6 Targeted lipidomics

The alternative to the global or shotgun approach to lipidomics is the targeted approach, where instead of trying to analyze all lipids at once, they are analyzed on a *class-by-class* basis. This is the approach adopted by the Lipid Maps consortium in the United States (http://www.lipidmaps.org/). To date, the collaborating groups have analyzed the macrophage-like cell line RAW 264.7 [16,17] and a pooled plasma sample representative of the U.S. population [15]. Six collaborating groups throughout the United States have analyzed the same sample but for different lipid classes (Figure 2.1). Each group has developed *class*-specific sample preparation and LC-ESI-MS/MS protocols (http://www.lipidmaps.org/downloads/2007_methods_chapters.pdf) [44], which they have then adopted in their analysis. The consortium was able to identify 229 distinct lipid species in the RAW 264.7 macrophage and >500 in plasma. The consortium also included a synthetic group to synthesize authentic standards, and as well as isotope-labeled standards for quantification, and a bioinformatics group to collate the data. A European consortium has also been established: Lipidomic Net (http://www.lipidomicnet.org/index.php/Main_Page).

2.7 Targeted steroidomics

Even when adopting a targeted approach, the coverage of lipids in a specific class can still be restricted by the dominance of a few very abundant lipids. This may be the case for steroid analysis where cholesterol is often the most abundant steroid by two orders of magnitude. This problem can be minimized by subdivision of the steroid class according to hydrophobicity where cholesterol and more hydrophobic steroids represent one class, and oxysterols (oxidized forms of cholesterol and its precursors) and less hydrophobic metabolites represent a second class [45]. This is an approach we adopted for the analysis of steroid sulphates and bile acid/ alcohol conjugates by shotgun steroidomics, and in our LC-ESI-MS/MS work on oxysterols and cholestenoic acids in brain [46,47], plasma [48], and cerebrospinal fluid (CSF) [49].

2.8 Oxysterols and cholestenoic acids

Oxysterols are oxidized forms of cholesterol: they are formed enzymatically in the first steps of cholesterol metabolism, and they may also be formed as a result of autoxidation of cholesterol [50]. The high abundance of cholesterol in biological systems means that oxysterols can easily be formed nonenzymatically during sample handling and workup unless care is taken. Oxysterols are biologically active molecules; they have been shown *in vitro* to activate nuclear receptors, for example, liver X receptors (LXRs) [51], to interact with INSIG protein and thereby to repress cholesterol synthesis [52], and, like bile acids, to interact with G-protein coupled receptors. Oxysterols also represent transport forms of cholesterol [53] (Table 2.3).

Oxysterols are abundant in mammalian brain (ng/mg) [45,46,54,55] (Table 2.3). They have been identified in both developing and adult brain [45–47], and there is strong evidence for their involvement in embryonic neuronal development [55,56]. Despite this, their analysis can be challenging. McDonald and colleagues have analyzed adult rodent brain using an LC-ESI-MS/MS methodology and identified the major sterols and oxysterols (Figure 2.10). However, to mine deeper into the brain steroidome, we have developed an analytical procedure specifically for sterols and steroids with a 3β-hydroxy-5-ene or 3-oxo-4-ene structure [46,47]. Brain tissue is homogenized in ethanol, and oxysterols separated from cholesterol and more hydrophobic sterols by SPE on a C_{18} column (Figure 2.11).

The oxysterol fraction is then treated with bacterial cholesterol oxidase, for example, from *streptomyces* sp., to convert 3β-hydroxy-5-ene functions to a 3-oxo-4-ene. The oxo groups are then reacted with the Girard P (GP) reagent, which effectively tags a positively charged quaternary nitrogen to the sterol structure. This process of *charge tagging* (Figure 2.12)

Table 2.3 Oxysterol Content of Rodent Brain and Human Cerebrospinal Fluid as Determined by Charge Tagging and LC-ESI-MS[n]

Oxysterol	Concentration in Wet Brain (ng/mg)[1]	Concentration in CSF[1] (ng/mL)	Comment/Enzymatic Formation[2]	References
24-Oxocholesterol	0.5 ng/mg (adult rat) 0.2 ng/mg (adult mouse) 0.03 ng/mg (neonatal mouse) 0.1 ng/mg (embryonic mouse)[3]		Isomerization product of 24S,25-epoxycholesterol	Karu, Hornshaw, Woffendin, et al. [46], Wang, Sousa, Bodin, et al. [47], Meljon, Griffiths, and Wang [57]
24S,25-Epoxycholesterol	0.1 ng/mg (neonatal brain) 0.03 ng/mg (embryonic mouse)[3]		Undergoes isomerization, hydrolysis, and methanolysis	Wang, Sousa, Bodin, et al. [47], Meljon, Griffiths, and Wang [57]
Total 24S,25-Epoxycholesterol[4]	0.6 ng/mg (adult rat) 0.4 ng/mg (adult mouse) 1.33 ng/mg (neonatal mouse) 0.33 ng/mg (embryonic mouse)[3]		Sum of 24S,25-epoxycholesterol, 24-oxocholesterol, 24,25-dihydroxycholesterol, 24-hydroxy,25-methoxycholesterol	Karu, Hornshaw, Woffendin, et al. [46], Wang, Sousa, Bodin, et al. [47], Meljon, Griffiths, and Wang [57]
24S-Hydroxycholesterol	15–25 ng/mg (adult rat) 30 ng/mg (adult mouse) 0.5 ng/mg (neonatal mouse) 0.03 ng/mg (embryonic mouse)[3]	0.02 ng/mL (human)	Cholesterol 24-hydroxylase (CYP46A1)[5]	Karu, Hornshaw, Woffendin, et al. [46], Wang, Sousa, Bodin, et al. [47], Ogundare, Theofilopoulos, Lockhart, et al. [49], Meljon, Griffiths, and Wang [57], Ogundare, Griffiths, Lockhart, et al. [58]
25-Hydroxycholesterol	0.6 ng/mg (adult mouse) 0.05 ng/mg (neonatal mouse) 0.01 ng/mg (embryonic mouse)[3]	0.04 ng/mL (human)	Cholesterol 25-hydroxylase (CH25H)[5]	Wang, Sousa, Bodin, et al. [47], Ogundare, Theofilopoulos, Lockhart, et al. [49], Meljon, Griffiths, and Wang [57], Ogundare, Griffiths, Lockhart, et al. [58]

26-Hydroxycholesterol	0.2 ng/mg (adult mouse) 0.01 ng/mg (neonatal mouse) 0.01 ng/mg (embryonic mouse)[3]	0.04 ng/mL (human)	Cholesterol 27-hydroxylase (CYP27A1)[5]	Wang, Sousa, Bodin, et al. [47], Ogundare, Theofilopoulos, Lockhart, et al. [49], Meljon, Griffiths, and Wang [57], Ogundare, Griffiths, Lockhart, et al. [58]
7α,25-Dihydroxycholesterol/ 7α,25-Dihydroxycholest-4-en-3-one[6]	0.03 ng/mg (adult rat)	0.01 ng/mL (human)	CH25H[5] followed by oxysterol 7α-hydroxylase (CYP7B1)/and hydroxysteroid dehydrogenase 3B7 (HSD3B7)	Karu, Hornshaw, Woffendin, et al. [46], Ogundare, Theofilopoulos, Lockhart, et al. [49], Ogundare, Griffiths, Lockhart, et al. [58]
7α,26-Dihydroxycholesterol/ 7α,26-Dihydroxycholest-4-en-3-one[6]	0.03 ng/mg (adult rat)	0.02 ng/mL (human)	CYP27A1[5] followed by CYP7B1/and hydroxysteroid dehydrogenase 3B7 (HSD3B7)	Karu, Hornshaw, Woffendin, et al. [46], Ogundare, Theofilopoulos, Lockhart, et al. [49], Ogundare, Griffiths, Lockhart, et al. [58]
24,25-Dihydroxycholesterol	0.1 ng/mg (adult rat) 0.1 ng/mg (adult mouse) 0.7 ng/mg (neonatal mouse) 0.1 ng/mg (embryonic mouse)[3]		May be endogenous or a hydrolysis product of 24,25-epoxycholesterol CYP46A1[5]	Karu, Hornshaw, Woffendin, et al. [46], Wang, Sousa, Bodin, et al. [47], Meljon, Griffiths, and Wang [57], Ogundare, Griffiths, Lockhart, et al. [58]
24-Hydroxy,25-methoxycholesterol[7]	0.1 ng/mg (adult mouse) 0.5 ng/mg (neonatal mouse) 0.1 ng/mg (embryonic mouse)[3]		Methanolysis products of 24,25-epoxycholesterol	Wang, Sousa, Bodin, et al. [47], Meljon, Griffiths, and Wang [57], Ogundare, Griffiths, Lockhart, et al. [58]

(continued)

Table 2.3 Oxysterol Content of Rodent Brain and Human Cerebrospinal Fluid a - Determined by Charge Tagging and LC-ESI-MS (Continued)[n]

Oxysterol	Concentration in Wet Brain (ng/mg)[1]	Concentration in CSF[1] (ng/mL)	Comment/Enzymatic Formation[2]	References
24,26-Dihydroxycholesterol	0.03 ng/mg (adult rat) 0.01 ng/mg (embryonic mouse)[3]		CYP46A1 followed by CYP27A1[5]	Karu, Hornshaw, Woffendin, et al. [46], Wang, Sousa, Bodin, et al. [47]
25,26-Dihydroxycholesterol	0.1 ng/mg (adult rat)		CH25H followed by CYP27A1[5]	Karu, Hornshaw, Woffendin, et al. [46]
3β-Hydroxycholest-5-en-26-oic acid		0.5 ng/mL (human)	Not profiled in brain studies CYP27A1	Ogundare, Theofilopoulos, Lockhart, et al. [49], Ogundare, Griffiths, Lockhart, et al. [58]
3β,7β-Dihydroxycholest-5-en-26-oic acid		0.3 ng/mL (human)	Not profiled in brain studies CYP27A1 followed by CYP7B1 and epimerase	Ogundare, Theofilopoulos, Lockhart, et al. [49], Ogundare, Griffiths, Lockhart, et al. [58]
7α-Hydroxy-3-oxocholest-4-en-26-oic acid		10 ng/mL (human)	Not profiled in brain studies CYP27A1 followed by CYP7B1 and HSD3B7	Ogundare, Theofilopoulos, Lockhart, et al. [49], Ogundare, Griffiths, Lockhart, et al. [58]
7α-Hydroxy-26-nor-cholest-4-ene-3,24-dione		0.3 ng/mL (human)	Not profiled in brain studies. Decarboxylation product of 7α-Hydroxy-3,24-bisoxocholest-4-en-26-oic acid CYP46A1 followed by CYP3A1, CYP27A1, HSD3B7, and L-bifunctional protein (LBP)	Ogundare, Theofilopoulos, Lockhart, et al. [49], Ogundare, Griffiths, Lockhart, et al. [58]

Compound	Concentration	Notes	Reference
7α-Hydroxy-3,24-bisoxocholest-4-en-26-oic acid	0.1 (human)	Not profiled in brain studies CYP46A1 followed by CYP39A1, CYP27A1, HSD3B7, and LBP	Ogundare, Theofilopoulos, Lockhart, et al. [49], Ogundare, Griffiths, Lockhart, et al. [58]
Total 7α-Hydroxy-3,24-bisoxocholest-4-en-26-oic acid[8]	0.4 ng/mL (human)	Sum of 7α-Hydroxy-26-nor-cholest-4-ene-3,24-dione 7α-Hydroxy-3,24-bisoxocholest-4-en-26-oic acid	Ogundare, Theofilopoulos, Lockhart, et al. [49], Ogundare, Griffiths, Lockhart, et al. [58]
7α,24-Dihydroxy-3-oxocholest-4-en-26-oic acid	2.1 ng/mL (human)	Not profiled in brain studies CYP46A1 followed by CYP39A1, CYP27A1, and HSD3B7	Ogundare, Theofilopoulos, Lockhart, et al. [49], Ogundare, Griffiths, Lockhart, et al. [58]
7α-Hydroxy-3-oxochol-4-en-24-oic acid	0.2 ng/mL (human)	CYP46A1 followed by CYP39A1, CYP27A1, HSD3B7, LBP, and sterol carrier protein x (SCPx)	Ogundare, Theofilopoulos, Lockhart, et al. [49], Ogundare, Griffiths, Lockhart, et al. [58]

1 Concentrations are for free unesterified sterols/acids.
2 Metabolic pathway depicted in Figure 2.17.
3 Embryonic mouse brain (E11).
4 Total 24S,25-epoxycholesterol calculated as the sum of 24S,25-epoxycholesterol, 24-oxocholesterol, 24,25-dihydroxycholesterol, and 24-hydroxy,25-methoxycholesterol.
5 As well as 24-hydroxylase activity, CYP46A1 also shows 25- and 26-hydroxylase activity. Mast, N. et al., 2003, Broad Substrate Specificity of Human Cytochrome P45046A1 Which Initiates Cholesterol Degradation in the Brain, *Biochemistry*. 42:14284–92.
6 Either dihydroxycholesterol or dihydroxycholest-4-en-3-one.
7 Either 24-hydroxy,25-methoxycholesterol and/or 24-methoxy,25-hydroxycholesterol.
8 Total 7α-hydroxy-3,24-bisoxocholest-4-en-26-oic acid sum of 7α-hydroxy-26-nor-cholest-4-ene-3,24-dione and 7α-hydroxy-3,24-bisoxocholest-4-en-26-oic acid.

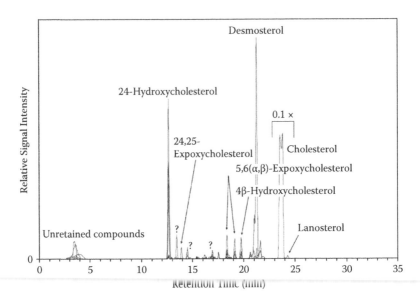

Figure 2.10 Reconstructed MRM traces for Bligh/Dyer extract of sterols from mouse brain. Known sterols are labeled, unknown compounds with a sterol-like signature are indicated by ?. MS/MS of the respective steroids can be found at Lipid Maps: http://www.lipidmaps.org/data/standards/standards.php?lipidclass=LMST. (Reproduced with permission from McDonald, J. G. et al., 2007, Extraction and Analysis of Sterols in Biological Matrices by High-Performance Liquid Chromatography Electrospray Ionization Mass Spectrometry, *Methods Enzymol.* 432:145–70.)

improves ESI sensitivity by two to three orders of magnitude and also adds specificity to the assay as only steroids with an oxo group, or oxidized by cholesterol oxidase to contain an one, will become derivatized. Analysis is then by LC-ESI-MS^n. A secondary advantage of charge tagging with the GP reagent is that it improves solubility of oxysterols in RP solvents, greatly increasing the diversity of RP mobile phases available for subsequent LC separations. The charged nature of the GP-tagged oxysterols dictates that they give intense signals in ESI spectra. When GP-tagged molecules fragment by MS/MS or MS^2 on tandem quadrupole, Q-TOF, or ion-trap instruments, the dominant fragment is the [M-79]$^+$ ion formed by loss of the pyridine group (Figures 2.13a and 2.14).

When analysis is performed on ion-trap instruments, this ion can be selected and subjected to further fragmentation. The resulting MS^3 spectra are rich in structural information (Figure 2.15). The MS/MS spectra on tandem quadrupole or Q-TOF instruments are effectively a composite of MS2 and MS3 spectra recorded on an ion trap. Using our charge-tagging methodology, we have profiled the oxysterol content of adult, newborn, and

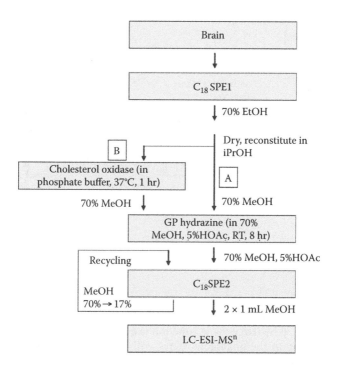

Figure 2.11 Extraction, charge tagging, and analysis of oxysterols. Oxysterols are extracted in ethanol, separated from cholesterol by dilution to 70% ethanol and passed through a C_{18} cartridge. Oxysterols appear in the flow-through, which is then dried and reconstituted in propan-2-ol and divided into fractions A and B. Fraction A is diluted to 70% methanol, 3% propan-2-ol, and 5% acetic acid and treated with excess GP reagent. Unreacted GP reagent is removed by a second C_{18} cartridge and charge-tagged oxysterols eluted with 100% methanol. Fraction B is treated in an identical fashion except oxysterols in propan-2-ol are treated with cholesterol oxidase in phosphaste buffer prior to dilution to 70% methanol. By analyzing fractions A and B by LC-ESI-MS, it is possible to differentiate oxysterols naturally containing an oxo group from those oxidized to contain one.

embryonic rodent brain. In adult brain, 24S-hydroxycholesterol (15–30 ng/mg) is the dominant oxysterol, and by using charge tagging with LC-ESI-MS^n analysis, we have also been able to identify 24,26-, 25,26-, 7α,25-, and 7α,26-dihydroxycholesterols (0.03–0.1 ng/mg) [46]. We also found 24-oxo-cholesterol (0.2–0.5 ng/mg) and 24,25-dihydroxycholesterol (0.1 ng/mg), which later studies revealed to be isomerization and hydrolysis products of 24S,25-epoxycholesterol, respectively (total 24S,25-epoxycholesterol in adult brain 0.4–0.6 ng/mg). 24S,25-Epoxycholesterol is formed via a shunt of the mevalonate pathway (Figure 2.16) and can be synthesized in parallel to cholesterol, although at a much lower rate, by the dual action of squalene

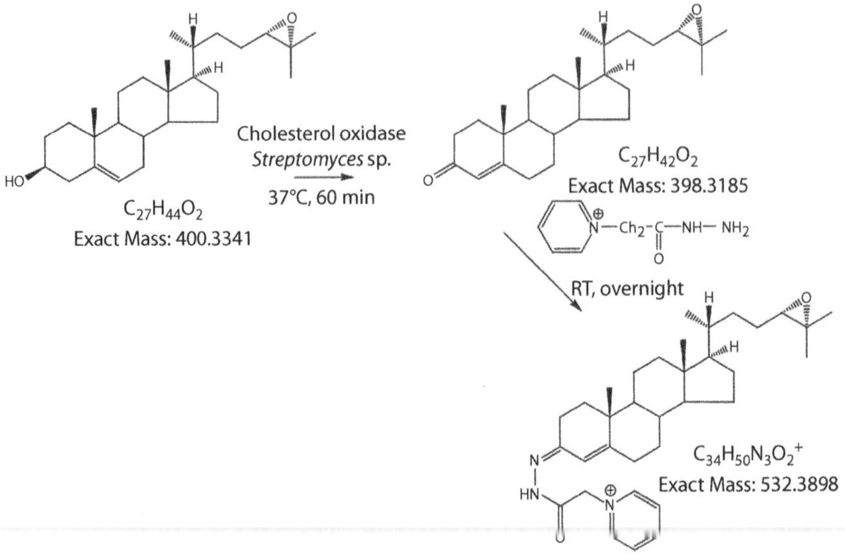

Figure 2.12 Charge tagging with the GP reagent to an oxysterol. Oxidation of an oxysterol with a 3β-hydroxy-5-ene function (e.g., 24S,25-epoxycholesterol, C^5-3β-ol-24S,25-epoxide) to the analogous sterol with a 3-oxo-4-ene function (24S,25-epoxycholest-4-en-3-one, C^4-24S,25-epoxy-3-one) followed by derivatization with GP reagent.

epoxidase, which introduces one or two oxygen atoms into squalene prior to cyclization to lanosterol and 24S,25-epoxylanosterol, and ultimately cholesterol and 24S,25-epoxycholesterol, respectively. Interestingly, in neonatal and embryonic brain (E11), the levels of 24S-hydroxycholesterol (0.5 ng/mg and 0.03 ng/mg, respectively) are much lower than in the adult [47,57]. This probably reflects the lack of cholesterol in a metabolically available pool during the early stages of life [55].

As in adult brain, 24-oxocholesterol and 24,25-dihydroxycholesterol are detected in neonatal and embryonic brain, as well as the 24S,25-epoxide itself and its methanolysis product 24-hydroxy,25-methoxy-cholesterol (and/or 24-methoxy,25-hydroxycholesterol) (see Table 2.3). The finding of high levels of total 24S,25-epoxycholesterol compared to 24S-hydroxycholesterol (see Table 2.3) in neonatal (1.33 ng/mg cf. 0.5 ng/mg) and embryonic (0.33 ng/mg cf. 0.03 ng/mg) brain is particularly interesting in light of a recent paper by Sacchetti et al. [56], which showed that oxysterols, such as 24S,25-epoxycholesterol, promote ventral midbrain neurogenesis through activation of LXRs which are expressed in developing brain, and that treatment of human embryonic stem cells with oxysterols increases neurogenesis and the number of dopaminergic neurons.

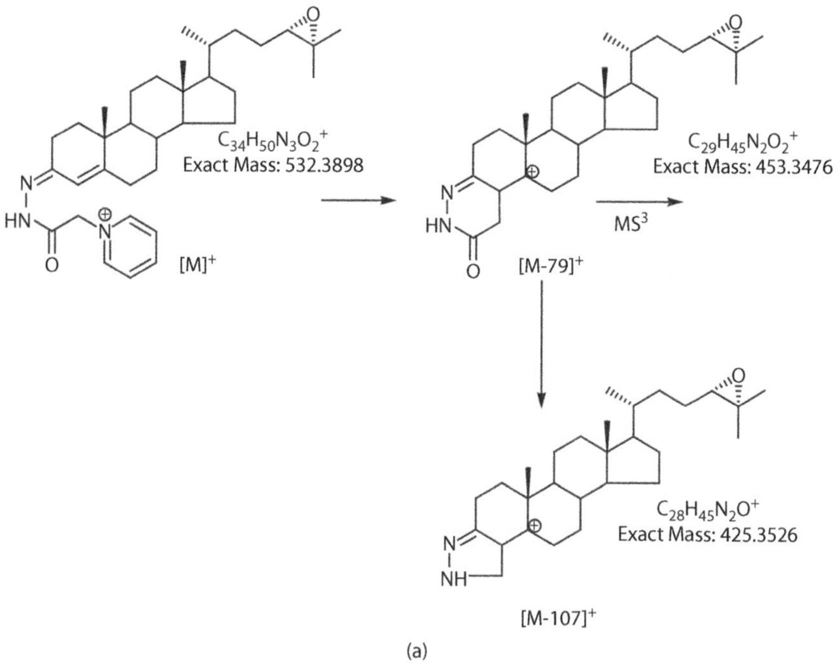

(a)

Figure 2.13a Fragmentation of GP-tagged oxysterols illustrated by 24S,25-epoxycholesterol. (a) Major fragments observed in MS^2 spectra recorded on an ion trap. (b) Major fragment observed in MS^3 spectra recorded on an ion trap. MS/MS spectra recorded on tandem quadrupole or Q-TOF instruments are a composite of (a) and (b). (c) The effect of 7α-hydroxylation on fragmentation as illustrated by 7α-hydroxycholesterol. (Continued)

Ventral midbrain is the region of embryonic brain from which dopaminergic neurons develop, and significantly, we were able to identify 24S,25-epoxycholesterol in this region of brain. Our results and those of Sacchetti et al. [56] suggest that LXR ligands, that is, 24S,25-epoxycholesterol, may have a role in stem cell replacement therapy for the treatment of Parkinson's disease, the hallmark of which is a loss of dopaminergic neurons.

In current work, we are using the above charge-tagging methodology to profile the oxysterol content of human cerebrospinal fluid to get a better view of the metabolism of cholesterol in living human brain [49,58]. We have been able to identify 12 different oxysterols, six C_{27} cholestenoic acids, and one C_{24} bile acid (Figure 2.17, Table 2.3). We are now testing the activity of these sterols and their precursors as ligands to nuclear receptors which are expressed in brain to investigate the signaling properties of these molecules in brain [49].

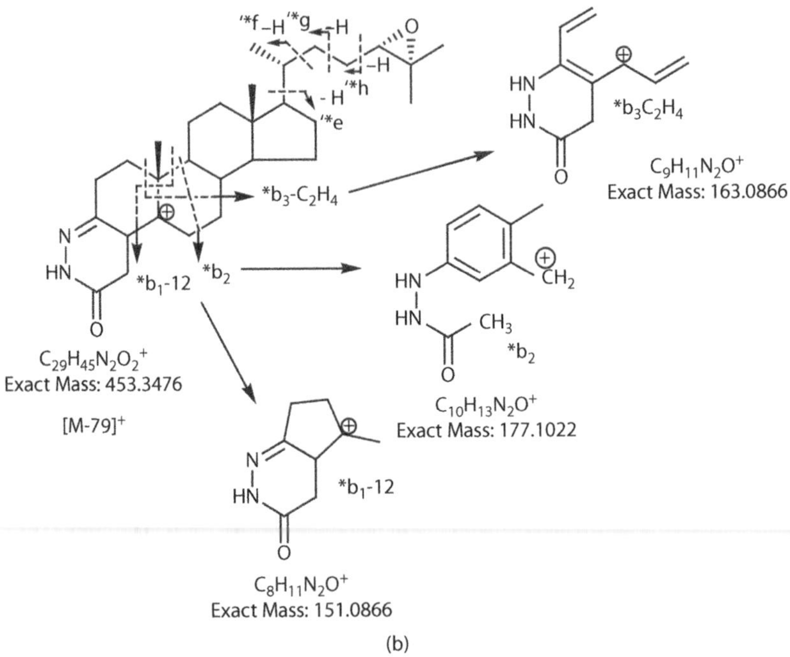

(b)

Figure 2.13b (Continued) Fragmentation of GP-tagged oxysterols illustrated by 24S,25-epoxycholesterol. (a) Major fragments observed in MS^2 spectra recorded on an ion trap. (b) Major fragment observed in MS^3 spectra recorded on an ion trap. MS/MS spectra recorded on tandem quadrupole or Q-TOF instruments are a composite of (a) and (b). (c) The effect of 7α-hydroxylation on fragmentation as illustrated by 7α-hydroxycholesterol. (Continued)

2.9 GC-MS analysis of oxysterols and cholestenoic and bile acids

GC-MS combines high-resolution gas-phase separations by gas–liquid chromatography with structurally informative and quantitative detection by EI-MS. A requirement of GC-MS is that the analyte is in the vapor phase and for steroid and bile acid analysis, this usually dictates that a derivatization step is necessary. The preferred derivatization of a carboxyl group is via methylation. This can be achieved by fresh diazomethane in ether added to a methanol solution of the sample. This method gives the highest specificity for carboxyl groups, and side reactions are few. Formation of methyl ethers is generally not a problem and is further prevented by low reaction temperature, limited concentrations of methanol, and short reaction times. Oxo groups may react to a minor extent with diazomethane. The drawback of using diazomethane is its toxicity in preparation and use, and this reagent requires care in handling [32].

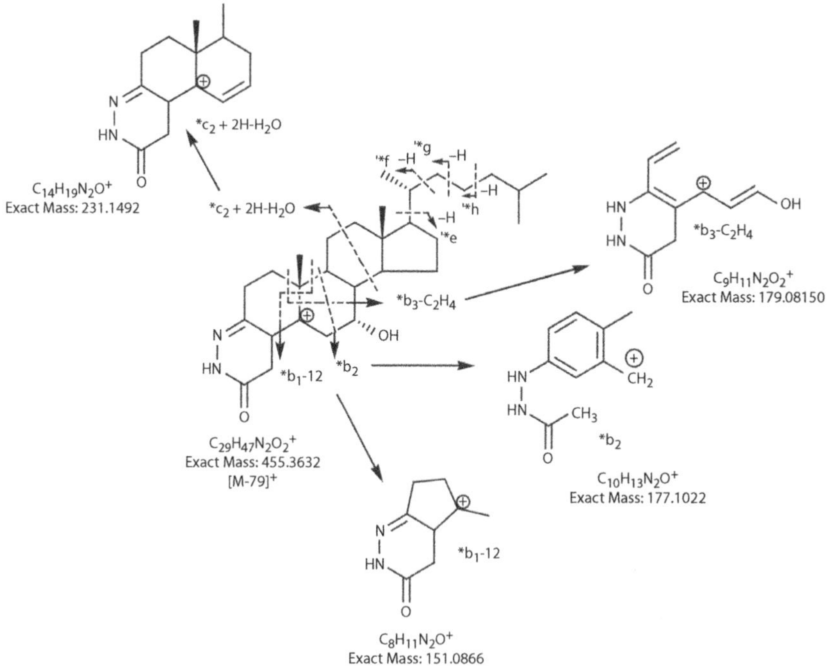

(c)

Figure 2.13c (Continued) Fragmentation of GP-tagged oxysterols illustrated by 24S,25-epoxycholesterol. (a) Major fragments observed in MS^2 spectra recorded on an ion trap. (b) Major fragment observed in MS^3 spectra recorded on an ion trap. MS/MS spectra recorded on tandem quadrupole or Q-TOF instruments are a composite of (a) and (b). (c) The effect of 7α-hydroxylation on fragmentation as illustrated by 7α-hydroxycholesterol.

An alternative reagent, trimethylsilyldiazomethane in hexane, is commercially available and appears useful for preparation of bile acid methyl esters [62]. Like carboxylic acid groups, hydroxyl groups also require protection prior to GC-MS analysis. A common and mild method to prepare trimethylsilyl (TMS) ethers is to react the sample with a mixture of dry pyridine, hexamethyldisilazane (HMDS), and trimethylchlorosilane (TMCS), 3:2:1 or 9:3:1 (by volume) with, or without heating [32]. However, if oxo groups are present in the bile acid, this reaction can yield enol-TMS ethers and multiple products. This artifact can be avoided by converting the oxo group into an oxime, usually a methyloxime (MO). To do this, the sample is dissolved in 50 μL pyridine with 5 mg methoxyammonium chloride and heated for 30 min at 60°C [32].

Oxysterols may exist in a free or conjugated form as esters with fatty and sulphuric acids and as an acetal with glucuronic acid. To analyze

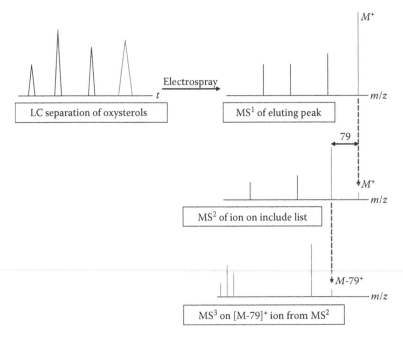

Figure 2.14 Schematic of an LC-ESI-MS³ experiment performed on an ion-trap instrument in the analysis of charge-tagged oxysterols.

total oxysterols by GC-MS, it is necessary to hydrolyze the esters prior to derivatization. This can be achieved by hydrolysis in ethanolic KOH, solvolysis with mineral acids, and hydrolysis with glucuronidase enzymes [32]. Bile acids can be similarly conjugated, but also at the carboxyl group with glycine and taurine; these amino acids are effectively hydrolyzed by commercially available microbial enzymes [32].

We have previously reviewed the GC-MS analysis of oxysterols and bile acids and the interested reader is directed to these works [32,33]. We believe that GC-MS offers a complimentary rather than competing analytical method to LC-MS/MS. It offers the major advantages of a reproducible retention index, separation of stereoisomers, reproducible and collated EI spectra, and the structurally rich EI spectra. The major disadvantage is the necessity to remove conjugating groups prior to GC-MS analysis, and the absence, or only minor abundance, of a molecular ion, thereby making definitive molecular weight determination difficult. In such situations, the analyst may resort to CI- rather than EI-MS. Although often seen as a disadvantage, derivatization also offers advantages as it adds chemical specificity to the assay and also may help direct EI fragmentation.

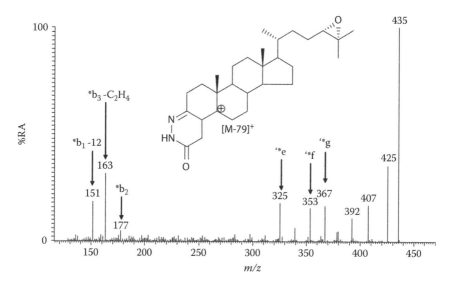

Figure 2.15 MS³ spectrum (532→453→) of GP-tagged 24S,25-epoxycholesterol. The spectrum was recorded on an LTQ Orbitrap XL. (Reproduced with permission from Wang, Y. et al., 2009, Targeted Lipidomic Analysis of Oxysterols in the Embryonic Central Nervous System, *Mol. Biosyst.* May:5:529–41.)

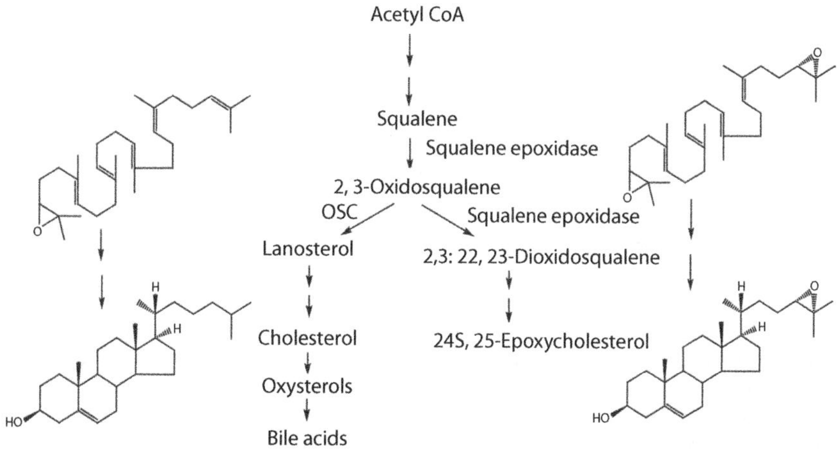

Acetyl CoA

↓

Squalene

Squalene epoxidase

2, 3-Oxidosqualene

OSC   Squalene epoxidase

Lanosterol   2,3: 22, 23-Dioxidosqualene

↓   ↓

Cholesterol   24S, 25-Epoxycholesterol

↓

Oxysterols

↓

Bile acids

Figure 2.16 Schematic view of the mevalonate pathway. Oxidosqualene cyclise (OSC).

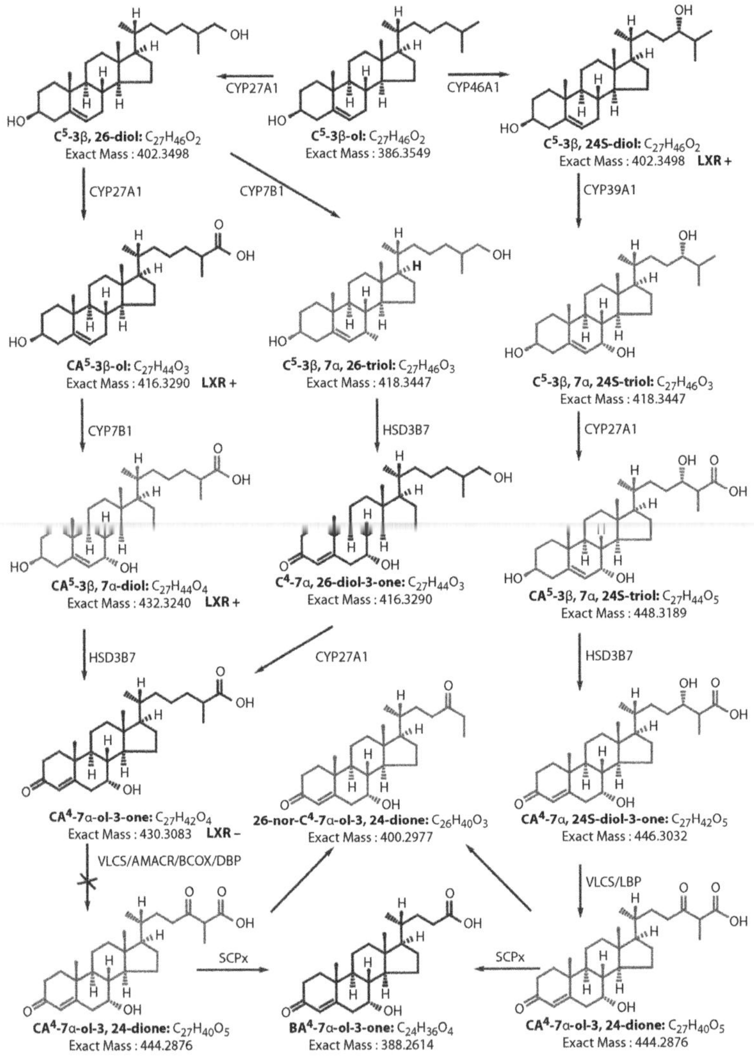

Figure 2.17 Suggested pathway of cholesterol metabolism in human brain based on the identification of metabolites in CSF. The left hand branch constitutes the initial steps of the acidic pathway of bile acid biosynthesis. 26-Hydroxycholesterol may be synthesized in brain via a reaction catalyzed by CYP27A1 or imported into brain from the circulation (Heverin, Meaney, Lütjohann, et al. [60].) Thus, the ultimate product of this pathway 7α-hydroxy-3-oxocholest-4-en-26-oic may be derived from brain synthesized or imported sterols. The right-hand branch of the pathway represents the initial steps of the 24-hydroxylase pathway of bile acid biosynthesis. The initial enzyme CYP46A1 is uniquely expressed in nervous tissue (Lund, Guileyardo, and Russell [61]). **(See color insert.)**

Table 2.4 Advantages/Disadvantages of Analytical Workflows in Lipidomics

Method	Advantage	Disadvantage
• Shotgun lipidomics	• Minimal sample handling • High throughput • Good sensitivity for easily ionized lipids • Semiquantitative	• Detection of minor components of complex mixtures restricted by instrument dynamic range • Discriminates against low abundance and poorly ionized lipids • Isomers not differentiated
• MDMS	• Ideal method for use on tandem quadrupole instruments • Utilizes precursor and neutral-loss scans to maximum effect	• When performed on quadrupole instruments, suffers from low resolution and low mass accuracy • When performed on Q-TOF or ion-trap instruments, time consuming
• Top-down lipidomics	• Exploits high-mass accuracy of high-resolution instruments • Incorporates structural information from MS/MS	• Can be time consuming if extensive MS/MS performed
• LC-MS/MS	• Sample complexity deconvoluted • Separation of isomers possible • Retention time adds an extra dimension to identification • Quantitative with use of stable isotope-labeled standards	• Reduced sample throughput • Dependent on reliability of chromatography system • MS/MS acquisition limited by chromatographic peak width

2.10 Conclusion

In this chapter, we have more or less provided an even-handed overview of the current status of lipidomics research. We have drawn examples of recent studies from our own work and others in the field. Others and we now adopt both a shotgun and also a targeted approach to lipidomics. Both strategies offer advantages (Table 2.4). Shotgun lipidomics is comparatively simple and offers a route toward high-throughput analysis. Targeted approaches based on LC-ESI-MS/ MS are more specialized and lower throughput but are more likely to provide a *gold standard* analysis. As lipidomics and metabolomics develop, metabolite databases are growing and searching possibilities based on exact mass are opening up. This will allow increased spectral annotation. But the discerning researcher is reminded that definitive identification can only be made by comparison to authentic standards. These may be of synthetic origin or isolated from natural products and fully characterized by LC-MS, GC-MS, and/or NMR. The importance of available standard compounds for definitive metabolite identification cannot be overstated.

Acknowledgments

This work was supported by the Biotechnology and Biological Sciences Research Council (BBSRC) and the Engineering and Physical Sciences Council (EPSRC), which are UK Research Councils.

References

1. Wang, Y. and Griffiths, W. J. 2008. Mass Spectrometry for Metabolite Identification. In *Metabolomics, Metabonomics and Metabolite Profiling*, ed. W. J. Griffiths, 1–43. Cambridge: Royal Society of Chemistry.
2. Viant, M. R., Ludwig, C., and Güther, U. L. 2008. 1D and 2D NMR Spectroscopy: From metabolic fingerprinting to profiling. In *Metabolomics, Metabonomics and Metabolite Profiling*, ed. W. J. Griffiths, 44–70. Cambridge: Royal Society of Chemistry.
3. Robinson, A. B. and Pauling, L. 1974. Techniques of orthomolecular diagnosis. *Clin. Chem.* 20:961–5.
4. Horning, E. C. 1968. Gas Phase Analytical Methods for the Study of Steroid Hormones and Their Metabolites. In *Gas Phase Chromatography of Steroids*, ed. K. B. Eik-Nes and H. C. Horning, 1–71. Berlin: Springer Verlag.
5. Beckmann, M., Parker, D., Enot, D. P., Duval, E., and Draper, J. 2008. High-throughput, nontargeted metabolite fingerprinting using nominal mass flow injection electrospray mass spectrometry. *Nat. Protoc.* 3:486–504.
6. Draper, J., Enot, D. P., Parker, D., Beckmann, M., Snowdon, S., Lin, W., and Zubair, H. 2009. Metabolite signal identification in accurate mass metabolomics data with MZedDB, an interactive m/z annotation tool utilizing predicted ionization behaviour "rules." *BMC Bioinformatics* 10:227.
7. Crews, B., Wikoff, W. R., Patti, G. J., Woo, H. K., Kalisiak, E., Heideker, J., and Siuzdak, G. 2009. Variability analysis of human plasma and cerebral spinal fluid reveals statistical significance of changes in mass spectrometry-based metabolomics data. *Anal. Chem.* 81:8538–44.
8. Wishart, D. S., Lewis, M. J., Morrissey, J. A., Flegel, M. D., Jeroncic, K., Xiong, Y., Cheng, D., Eisner, R., Gautam, B., Tzur, D., Sawhney, S., Bamforth, F., Greiner, R., and Li, L. 2008. The human cerebrospinal fluid metabolome. *J. Chromatogr. B* 871:164–73.

9. Dunn, W. B., Broadhurst, D., Brown, M., Baker, P. N., Redman, C. W., Kenny, L. C., and Kell, D. B. 2008. Metabolic profiling of serum using ultra performance liquid chromatography and the LTQ-orbitrap mass spectrometry system. *J. Chromatogr. B* 871:288–98.

10. Blau, K. and Halket, J. M. 1993. *Handbook of Derivatives for Chromatography*, 2nd ed. Chichester: Wiley.

11. Cech, N. B. and Enke, C. G. 2001. Practical implications of some recent studies in electrospray ionization fundamentals. *Mass Spectrom. Rev.* 20:362–387.

12. Dennis, E. A. 2009. Lipidomics joins theomics evolution. *Proc. Natl. Acad. Sci. USA* 106:2089–90.

13. Shevchenko, A. and Simons, K. 2010. Lipidomics: Coming to grips with lipid diversity. *Nat. Rev. Mol. Cell Biol.* 11:593–8.

14. Griffiths, W. J. and Wang, Y. 2009. Mass spectrometry: From proteomics to metabolomics and lipidomics. *Chem. Soc. Rev.* 38:1882–96.

15. Quehenberger, O., Armando, A. M., Brown, A. H., Milne, S. B., Myers, D. S., Merrill, A. H., Bandyopadhyay, S., Jones, K. N., Kelly, S., Shaner, R. L., et al. 2010. Lipidomics reveals a remarkable diversity of lipids in human plasma. *J. Lipid Res.* 51:3299–305.

16. Andreyev, A. Y., Fahy, E., Guan, Z., Kelly, S., Li, X., McDonald, J. G., Milne, S., Myers, D., Park, H., Ryan, A., et al. 2010. Subcellular organelle lipidomics in TLR-4-activated macrophages. *J. Lipid Res.* 51:2785–97.

17. Dennis, E. A., Deems, R. A., Harkewicz, R., Quehenberger, O., Brown, H. A., Milne, S. B., Myers, D. S., Glass, C. K., et al. 2010. A mouse macrophage lipidome. *J. Biol. Chem.* Published on October 5, 2010 as Manuscript M110.182915, 285:39976–85.

18. Graessler, J., Schwudke, D., Schwarz, P. E., Herzog, R., Shevchenko, A., and Bornstein, S. R. 2009. Top-down lipidomics reveals ether lipid deficiency in blood plasma of hypertensive patients. *PLoS One* 4:e6261.

19. Han, X. 2010. Multi-dimensional mass spectrometry-based shotgun lipidomics and the altered lipids at the mild cognitive impairment stage of Alzheimer's disease. *Biochim. Biophys. Acta* 1801:774–83.

20. Han, X. and Gross, R. W. 2003. Global analyzes of cellular lipidomes directly from crude extracts of biological samples by ESI mass spectrometry: A bridge to lipidomics. *J. Lipid Res.* 44:1071–9.

21. Han, X. and Gross, R. W. 2005. Shotgun lipidomics: Electrospray ionization mass spectrometric analysis and quantitation of cellular lipidomes directly from crude extracts of biological samples. *Mass Spectrom. Rev.* 24:367–412.

22. Han, X., Yang, K., Cheng, H., Fikes, K. N., and Gross, R. W. 2005. Shotgun lipidomics of phosphoethanolamine-containing lipids in biological samples after one-step *in situ* derivatization. *J. Lipid Res.* 46:1548–60.

23. Yang, K., Cheng, H., Gross, R. W., and Han, X. 2009. Automated lipid identification and quantification by multidimensional mass spectrometry-based shotgun lipidomics. *Anal. Chem.* 81:4356–68.

24. Schwudke, D., Hannich, J. T., Surendranath, V., Grimard, V., Moehring, T., Burton, L., Kurzchalia, T., and Shevchenko, A. 2007. Top-down lipidomic screens by multivariate analysis of high-resolution survey mass spectra. *Anal. Chem.* 79:4083–93.

25. Christie, W. W. and Han, X. 2010. *Lipid Analysis—Isolation, Separation, Identification and Lipidomic Analysis*. Bridgewater, England: The Oily Press Lipid Library.

26. Hunt, A. N., Macken, M., Koster, G., Kohler, J. A., and Postle, A. D. 2008. Diclofenac mediated derangement of neuroblastoma cell lipidomic profiles is accompanied by increased phosphatidylcholine biosynthesis. *Adv. Enzyme Regul.* 48:74–87.

27. Ejsing, C. S., Duchoslav, E., Sampaio, J., Simons, K., Bonner, R., Thiele, C., Ekroos, K., and Shevchenko, A. 2006. Automated identification and quantification of glycerol phospho lipid molecular species by multiple precursor ion scanning. *Anal. Chem.* 78:6202–14.

28. Fahy, E., Subramaniam, S., Brown, H. A., Glass, C. K., Merrill, A. H. Jr., Murphy, R. C., Raetz, C. R., Russell, D. W., Seyama, Y., Shaw, W., Shimizu, T., Spener, F., Van Meer, G., Van Nieuwenhze, M. S., White, S. H., Witztum, J. L., and Dennis, E. A. 2005. A comprehensive classification system for lipids. *J. Lipid Res.* 46:839–61.

29. Griffiths, W. J. 2003. Tandem mass spectrometry in the study of fatty acids, bile acids, and steroids. *Mass Spectrom. Rev.* 22:81–152

30. Pulfer, M. and Murphy, R. C. 2003. Electro spray mass spectrometry of phospholipids. *Mass Spectrom. Rev.* 22:332–64.

31. Beynon, J. H. 1954. Qualitative analysis of organic compounds by mass spectrometry. *Nature* 174:735.

32. Sjövall, J., Griffiths, W. J., Setchell, K. D., Mano, N., and Goto, J. 2010. Analysis of Bile Acids. In *Steroid Analysis,* 2nd ed., ed. H. L. Makin and D. B. Gower. Springer.

33. Makin, H. L., Honour, J. E., Shackleton, C. H., and Griffiths, W. J. 2010. General Methods for the Extraction, Purification, and Measurement of Steroids by Chromatography and Mass Spectrometry. In *Steroid Analysis,* 2nd ed., ed. H. L. Makin and D. B. Gower. New York: Springer.

34. Shackleton, C. H. and Whitney, J. O. 1980. Use of Sep-Pak cartridges for urinary steroid extraction: Evaluation of the method for use prior to gas chromatographic analysis. *Clin. Chim. Acta* 107:231–243.

35. Shackleton, C. H., Merdinck, J., and Lawson, A. M. 1990. Steroid and Bile Acid Analyzes. In *Mass Spectrometry of Biological Materials,* ed. C. N. McEwen and B. S. Larsen, 297–377. New York: Marcel Dekker.

36. Shackleton, C. H., Kletke, C., Wudy, S., and Pratt, J. H. 1990. Dehydroepiandrosterone sulfate quantification in serum using high-performance liquid chromatography/mass spectrometry and a deuterated internal standard: A technique suitable for routine use or as a reference method. *Steroids* 55:472–78.

37. Nordström, A., Want, E., Northen, T., Lehtiö, J., and Siuzdak, G. 2008. Multiple ionization mass spectrometry strategy used to reveal the complexity of metabolomics. *Anal. Chem.* 80:421–429.

38. Liu, S., Griffiths, W. J. and Sjövall, J. 2003. Capillary liquid chromatography/electrospray mass spectrometry for analysis of steroid sulfates in biological samples. *Anal. Chem.* 75:791–7.

39. Setchell, K. D. and Heubi, J. E. 2006. Defects in bile acid biosynthesis—Diagnosis and treatment. *J Pediatr. Gastroenterol. Nutr.* 43 Suppl 1:S17–22.

40. Schiller, J., Suss, R., Fuchs, B., Muller, M., Zschornig, O., and Arnold, K. 2007. MALDI-TOFMS in lipidomics. *Front. Biosci.* 12:2568–79.

41. Wiseman, J. M., Puolitaival, S. M., Takats, Z., Cooks, R. G., and Caprioli, R. M. 2005. Mass spectrometric profiling of intact biological tissue by using desorption electrospray ionization. *Angew. Chem. Int. Ed. Engl.* 44:7094–7.

42. Jiye, A., Trygg, J., Gullberg, J., Johansson, A. I., Jonsson, P., Antti, H., Marklund, S. L., and Moritz, T. 2005. Extraction and GC/MS analysis of the human blood plasma metabolome. *Anal. Chem.* 77:8086–94.

43. Dunn, W. B., Broadhurst, D., Ellis, D. I., Brown, M., Halsall, A., O'Hagan, S., Spasic, I., Tseng, A., and Kell, D. B. 2008. A GC-TOF-MS study of the stability of serum and urine metabolomes during the UK biobank sample collection and preparation protocols. *Int. J. Epidemiol.* 37(Suppl. 1):i23–i30.

44. Schmelzer, K., Fahy, E., Subramaniam, S., and Dennis, E. A. 2007. The lipid maps initiative in lipidomics. *Methods Enzymol.* 432:171–83.

45. McDonald, J. G., Thompson, B. M., McCrum, E. C., and Russell, D. W. 2007. Extraction and analysis of sterols in biological matrices by high performance liquid chromatography electrospray ionization mass spectrometry. *Methods Enzymol.* 432:145–70.

46. Karu, K., Hornshaw, M., Woffendin, G., Bodin, K., Hamberg, M., Alvelius, G., Sjövall, J., Turton, J., Wang, Y., and Griffiths, W. J. 2007. Liquid chromatography-mass spectrometry utilizing multi-stage fragmentation for the identification of oxysterols. *J. Lipid Res.* 48:976–87.

47. Wang, Y., Sousa, K. M., Bodin, K., Theofilopoulos, S., Sacchetti, P., Hornshaw, M., Woffendin, G., Karu, K., Sjövall, J., Arenas, E., and Griffiths, W. J. 2009. Targeted lipidomic analysis of oxysterols in the embryonic central nervous system. *Mol. Biosyst.* 5:529–41.

48. Griffiths, W. J., Hornshaw, M., Woffendin, G., Baker, S. F., Lockhart, A., Heidelberger, S., Gustafsson, M., Sjövall, J., and Wang, Y. 2008. Discovering oxysterols in plasma: A window on the metabolome. *J. Proteome Res.* 7:3602–12.

49. Ogundare, M., Theofilopoulos, S., Lockhart, A., Hall, L. J., Arenas, E., Sjövall, J., Brenton, A. G., Wang, Y., and Griffiths, W. J. 2010. Cerebrospinal fluid steroidomics: Are bioactive bile acids present in brain? *J. Biol. Chem.* 285:4666–79.

50. Schroepfer, G. J. Jr. 2000. Oxysterols: Modulators of cholesterol metabolism and other processes. *Physiol. Rev.* 80:361–554.

51. Janowski, B. A., Grogan, M. J., Jones, S. A., Wisely, G. B., Kliewer, S. A., Corey, E. J., and Mangelsdorf, D. J. 1999. Structural requirements of ligands for the oxysterolliver X receptors LXR alpha and LXR beta. *Proc. Natl. Acad. Sci. USA* 96:266–71.

52. Radhakrishnan, A., Ikeda, Y., Kwon, H. J., Brown, M. S., and Goldstein, J. L. 2007. Sterol-regulated transport of SREBPs from endoplasmicreticulum to Golgi: Oxysterols block transport by binding to insig. *Proc. Natl. Acad, Sci, USA* 104:6511–18.

53. Björkhem, I. and Diczfalusy, U. 2002. Oxysterols: Friends, foes, or just fellow passengers? *Arterioscler. Thromb. Vasc. Biol.* 22:734–42.

54. Lütjohann, D., Breuer, O., Ahlborg, G., Nennesmo, I., Sidén, A., Diczfalusy, U., and Björkhem, I. 1996. Cholesterol homeostasis in human brain: Evidence of an age-dependent flux of 24S-hydroxycholesterol from the brain into the circulation. *Proc. Natl. Acad. Sci. USA* 93:9799–804.

55. Dietschy, J. M. and Turley, S. D. 2004. Thematic review series: Brain lipids. Cholesterol metabolism in the central nervous system during early development and in the mature animal. *J. Lipid Res.* 45:1375–97.

56. Sacchetti, P., Sousa, K. M., Hall, A. C., Liste, I., Steffensen, K. R., Theofilopoulos, S., Parish, C. L., Hazenberg, C., Richter, L. Ä., Hovatta, O., Gustafsson, J. Å., and Arenas, E. 2009. Liver X receptors and oxysterols promote ventral midbrain neurogenesis *in vivo* and in human embryonic stem cells. *Cell. Stem Cell* 5:409–19.

57. Meljon, A., Griffiths, W. J., and Wang, Y. 2010. Analysis of Oxysterols in Mature and Developing Brain, Charge-Tagging and LC-MS^n. Paper presented at the annual meeting of the British Mass Spectrometry Society, Cardiff.

58. Ogundare, M., Griffiths, W. J., Lockhart, A., Sjövall, J., and Wang, Y. 2010. Analysis of Oxysterols and Cholestenoic Acids in Plasma and CSF by Charge-Tagging and LC-MS^n. Paper presented at the annual meeting of the British Mass Spectrometry Society, Cardiff.

59. Mast, N., Norcross, R., Andersson, U., Shou, M., Nakayama, K., Björkhem, I., and Pikuleva, I. A. 2003. Broad substrate specificity of human cytochrome P45046A1 which initiates cholesterol degradation in the brain. *Biochemistry* 42:14284–92.

60. Heverin, M., Meaney, S., Lütjohann, D., Diczfalusy, U., Wahren, J., and Björkhem, I. 2005. Crossing the barrier: Net flux of 27-hydroxycholesterol into the human brain. *J. Lipid Res.* 46:1047–52.

61. Lund, E. G., Guileyardo, J. M., and Russell, D. W. 1999. cDNA cloning of cholesterol 24-hydroxylase, a mediator of cholesterol homeostasis in the brain. *Proc. Natl. Acad. Sci. USA* 96:7238–43.

62. Alvelius, G., Hjalmarson, O., Griffiths, W. J., Björkhem, I., and Sjövall, J. 2001. Identification of unusual 7-oxygenated bile acid sulfates in a patient with Niemann-Pick disease, type C. *J. Lipid Res.* 42:1571–7.

chapter three

From chemical genomics to chemical proteomics
The power of microarray technology

Sándor Spisák and András Guttman

Contents

3.1   Introduction

Microarray technology is a high-throughput screening tool to monitor gene and protein expression-level changes and to study their interactions with other biologically important molecules [1–6]. The advent of nucleic acid and protein microarrays is a major advancement in contemporary chemical genomics and chemical proteomics, including drug discovery, medicinal chemistry, and clinical diagnostics. DNA microarrays with up to tens of thousands of oligonucleotide probes are used to measure changes in gene expression levels, to detect single nucleotide polymorphisms (SNPs) and copy number variations (CNVs), or for genotyping or resequencing. In protein microchip technology, in most instances, planar antibody arrays are used with immobilized, monoclonal, polyclonal, or monospecific antibodies to capture their respective antigens [7–9]. Fluorescent or chemiluminescent methods are used for detection and quantification [10–12]. Another approach of chemical proteomics is the use of reversed phase protein microarrays, where proteins are bound to a solid surface, and antibodies from cell lysates or protein extracts [13–17] are used for their interrogation. Whole-cell arrays utilize immobilized cell surface antigen-specific antibodies, which exclusively bind the corresponding cells, allowing detection of surface proteins of interest without their isolation [18,19]. Other techniques utilize suspension or microbead-based arrays, and more recently, affinity and/or chromatographic surface arrays [20–25]. Protein expression profiling using microarray technology is a logical continuation of gene expression profiling (cDNA and transcriptome), with the goal to shed light on the actual expression products of genes that play roles in specific biological processes, diseases, and responses to drug treatments [26–28]. cDNA microarrays are extensively used to identify gene expression changes, as well as to follow expression patterns at the genome level and in comparative phenotype and/or physiological state studies [29]. However, in most instances, cDNA microarrays do not provide relevant information about protein expression levels, that is, one can draw some conclusion about the properties of the inherited material (mutation, deletion, addition, repeats, translocation, etc.) at the transcriptome level, but apparently no information about their gene products, that is, the proteins

[30], and much less about their posttranslational modifications. Indeed, except for some rare cases [31], large differences have been reported between transcriptome and proteome expression levels within the same sample [9,32,33].

Protein microarrays combine the advantages of cDNA microarray technology and Western blotting. From the detection point of view, while cDNA microarrays are evaluated by DNA (Southern) or RNA (Northern) hybridization methods, interrogation of protein microchips is done by modified Western-type hybridization techniques. Here the antibody (or antigen, in the case of reverse phase array) (RPA) probe is in a fixed position on the microarray surface, enabling simultaneous interrogation of up to a few thousand sample components. Comparative profiling and multiplexing is possible by the utilization of different dye-labeled samples (normal versus diseased or treated versus untreated, etc.) [8,34]. In this chapter, we summarize the latest developments in microarray technologies utilized in chemical genomics and chemical proteomics to simultaneously study a large number of samples in order to discover novel biomarkers.

3.1.1 High-throughput molecular biology approaches in omics applications

To understand complex genetic diseases, many biological parameters should be simultaneously interrogated by high-throughput methods. The Human Genome Project in the early 90s made possible the widely spread use of microarray technology, suitable for accomplishing hundreds or thousands of tests at the same time [35]. The advent of cDNA microarray technology led to revolutionary changes in gene expression studies. Introduction of high-density oligonucleotide microarrays enabled transcriptome profiling at the whole-genome level. Recent developments in next-generation sequencing systems (NGS) extended this opportunity even further as previously unknown splice variants can now be quantitatively identified [36]. At the proteome level, in addition to the very important mass spectrometry (MS)-based techniques, advances in microarray technology made protein chips available to examine the target phenotype. Other types of microarrays, such as carbohydrate, lipid, and aptamer chips, are also available to screen posttranslational modifications and metabolic processes. In the following paragraphs, the theoretical basis for microarray technology, the operating principles, and the main application areas to be described also critically point out the limitations of the method. Genomics research provided information on the molecular biology basis of diseases and recognized genes that play important roles in biological processes related to drug treatments. Our knowledge about these genes is continuously growing by utilizing modern mRNA technologies, including Serial

Analysis of Gene Expression (SAGE), microarray methods, and next-generation sequencing (NGS) [35,37,38]. Complementary DNA (cDNA) microarrays can be used to measure changes in gene expression levels and to evaluate gene expression patterns in comparative studies with different phenotypes or physiological conditions. However, as emphasized earlier, DNA microchips are not capable of providing information about changes at the proteome level [39].

3.2   Transcriptome analysis

3.2.1   Microarray-based methods

3.2.1.1   History of microarray technology

Microarray technology has evolved considerably since its introduction approximately two decades ago and has brought a major breakthrough in molecular biology making possible research at the biomolecular level (molecular diagnostics). Early attempts were analogous to Southern blotting technology, using a so-called dot blot and gridded library approach employing large membranes and radioactive isotope labeling. With the advent of special printing needles, fabrication of much higher-density arrays were possible, such as 100 elements/cm^2 on nonporous glass slides [40]. With this approach, the reaction volume was significantly reduced and the reaction rate was increased. The use of fluorescent dyes, however, sometimes caused strong nonspecific background on porous membrane. The novel cyanine dyes (Cy3, Cy5) provided excellent signal intensity, allowing quantitative evaluation with the option of using two or more fluorophores at the same time, making possible simultaneous comparative study of two or more samples in one reaction (slide). Unfortunately, this approach did not become widespread because these two dyes (Cy3, Cy5) had different bonding abilities to the target molecule and required different excitation wavelengths for their detection.

The basis of contemporary microarray technology is printing known oligonucleotide sequences in predefined positions onto the array surface, which then capture the labeled complementary strands from the sample, following the complementarity rules (duplex forming and hybridization). After hybridization, the original amount of mRNA molecules can be estimated from the signal intensity of the captured/hybridized molecules. In the case of the microscope slide-based microarrays (also referred to as *cDNA microarray*), presynthesized complementary DNA molecules are attached to a solid surface [40].

3.2.1.2   Microarray types

DNA microchips are categorized based on their structure and function (Table 3.1). Literature reports mostly refer to chip capacities of detecting

Table 3.1 Most Popular Microarray-Based Techniques

Test Method	Sample Amount	Cost	Number of Transcripts	Advantages	Disadvantages	Comment
cDNA microarray	10–100 µg	Small	100–10,000	Costs, appropriate for screening, open system, flexible	Low reproducibility, time-consuming, spots inhomogeneity, high background/labeling effect	Mainly used for gene expression studies, microscope glass slide–based approaches
Two-color oligonucleotide microarray	1–10 µg	High	1000–50,000	Well established, automated, wide applicability	Competition between the two samples, different fluorescence intensity per channel/dye	Ink jet technology, high density, gene expression analysis, DNA sequencing, SNP analysis, cytogenetics, CGH, CNV, ChIP, miRNA
One-channel oligonucleotide microarray	100 ng–10 µg 50 ng–500 ng	Very high	10,000–50,000	Well established, automated, sensitive, specific, very high density, small sample volume, wide applicability, very high resolution	Labor-intensive, time-consuming, necessity of biocomputing, closed platform, annotation error possibility	Photolithography, laser micromirror technology, indirect labeling technology, gene expression analysis, DNA sequencing, SNP analysis, cytogenetic, CGH, CNV, ChIP, miRNA
Bead-based array	Different, optimization needed	Small/ modest	1–100	Quick and high sensitivity	Labor-intensive	Highly recommeded screening for large sample number with few interests, mutation analysis, SNP, gene expression, miRNA

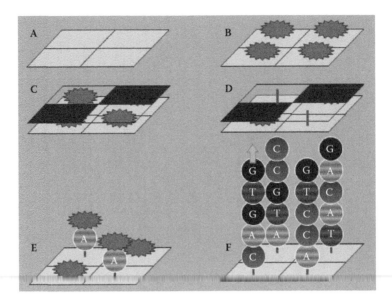

Figure 3.1 Schematic representation of solid phase oligonucleotide synthesis. (A) Glass chip surface. (B) Conjugation of light-sensitive groups. (C) Photomasking. (D) Formation of reactive groups. (E) Binding the first set of nucleotides. (F) Synthesized probes.

1000 to 10,000 target molecules. The molecules fixed to the solid surface are referred to as *probes* (oligonucleotides, cDNA, polymerase chain reaction (PCR) products, or chromosome fragments), to which the target molecules are hybridized. Microarrays are mostly made of glass, nylon, or plastic. In DNA chips, probe cDNA sequences and shorter oligonucleotides are often used, but even entire denatured chromosomes can be immobilized. Based on the way they are fabricated (printed), we distinguish *on-chip* and *off-chip* technologies. In the on-chip method, *in situ* oligonucleotide synthesis is applied with the help of photolithography (Affymetrix®) (Figure 3.1), maskless micromirror (NimbleGen), and inkjet (SurePrint, Agilent) technology, while in the off-chip method, contact and noncontact techniques are applied [41–45]. Furthermore, there are low-density (100 to 1000 target-containing) *active* chips, primarily used in diagnostics. The term *active* means that the probes are being synthesized by active manner, such as photolithography, inkjet technology, and so forth, to the wafer. Medium-target density chips (1000 to 10,000 probes) are mostly used for gene expression pattern analysis (drug targets, disease-specific array), while high-density chips (100,000 or more probes) are used for global polymorphism studies. Chips containing millions of features are suitable for whole-genome transcript and/or splice variant analysis. They are also called *large complexity–high-density microarrays*. The labeling methods can be radioactive isotope (P32, P33), fluorescent (Cy3, Cy5 or Streptavidin, R-phycoerythrin conjugate [SAPE]), and chemiluminescent (biotinylated nucleotides + Streptavidin-Alkaline phosphatase conjugate [SA-AP] + substrate) based. Detection is accomplished accordingly, that is, by laser scanner of phosphor image analyzer systems [40].

3.2.1.3   Oligonucleotide microarrays: On-chip technology

Photolithography is a well-established method in the microelectronics industry, which made possible fabrication of very-high-density micro-chips. Oligonucleotides are readily synthesized onto a solid wafer sur-face by the help of photosensitive groups. The technology uses 25-mers, which are produced using 100 (4 × n, n = 25) photolithography masks. The oligonucleotides obtained in this way are referred to as *probes* and their density can be as high as 1,000,000 per spot. Each transcript is rep-resented by eleven 25-mer oligonucleotides probes, which are scattered around on the chip surface. They are originated from the 3′ end of the relevant mRNA and covered (evenly or overlapping) an approximately 1000 bp region. Therefore, only a specific section (complementary part) of the complete cDNA hybridizes to the probe. After the hybridization reaction, the fluorescence intensity is measured at all 11 positions and the signal is processed by a special algorithm. These intensity values are stored and the data can be obtained by fitting a grid to the hybridization control points. Each cell consists of 7 × 7 pixels, but to eliminate the risk of fitting errors with the array grid, only a smaller pixel layer of 5 × 5 is evaluated. Proper fitting of the grid is a challenge, because if the chip sur-face is not uniform or the matrix (1200 × 1200) is distorted by as little as only 0.01 degrees, the error in matching can be significant. The raw fluo-rescence data is stored in a cell file (CEL) that corresponds to each pixel position of the measured fluorescence intensity values on the scanned image (Figure 3.2). Evaluation of the microarrays is performed after the preprocessing steps (background correction, normalization, and summa-rization) by multivariate statistical analysis using CEL files. The two basic approaches are differential expression analysis and prediction analysis. Differential expression analysis identifies genes that are differentially expressed among the groups examined, thus this process is capable of identifying down- and/or up-regulated genes. Prediction analysis aims to identify gene expression pattern differences in order to recognize various groups [46,47].

In addition to gene expression studies, microarray technology made it possible to perform other DNA-level studies, such as DNA sequencing (resequencing array) [48–52], SNP identification, genotyping (SNP array) [53–55], karyotyping, cytological studies, and comparative genomic hybridization (CGH) [56,57]. Resequencing arrays have the largest (maxi-mal) resolution (1 bp), which means that the overlapped probes are shifted

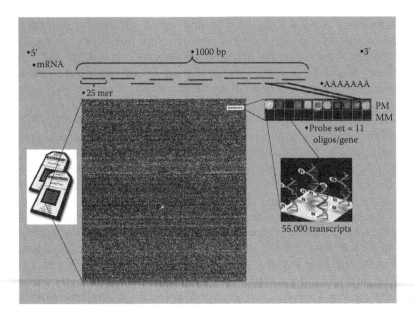

Figure 3.2 Design of a whole-genome microarray. Fixing the probes (25-mers) to the wafer surface. Probes for one gene are scattered on the chip surface. Eleven probes form one unit called a *probe set*. In the upper right section, a probe set is shown containing 11 perfectly homologous, *perfect* match (PM), probes and 11 control, *mismatch* (MM), probes not containing complementary bases at position 13.

by one base on the target molecule allowing sequence identification. Non-overlapping probes are suitable for SNP analysis and cytological studies. There are possibilities for performing gene regulation studies through transcription factor, binding-site determination, methylation analysis (ChIP-on-chip) [58–60], or via microRNA (miRNA) identification. For these latter applications, entire genomes, or pieces of DNA (complementary to miRNA sequences), are fixed to the chip surface [61]; however, with the advent of new generation sequencing methods, their importance is continuously decreasing due to their high costs, the large amount of sample required, and possible hybridization errors.

3.3   Genome analysis

3.3.1   DNA sequencing

DNA sequencing is a method for determining the order of bases within a DNA molecule, that is, adenine, cytosine, guanine, and thymine. DNA sequencing has been one of the main focuses of technological development since Nobel laureates Sanger and Gilbert introduced sequencing by chain-termination and chemical-fragmentation techniques coupled with gel electrophoresis-based size separation [35,37]. The key principle of the Sanger method, the use of dideoxynucleotide triphosphates as DNA chain terminators, proved to be more efficient, requiring fewer toxic chemicals and lower amounts of radioactivity than that of the Maxam–Gilbert chemical-fragmentation method. The classical chain-termination method required a DNA primer (radioactively or fluorescently labeled), a single-stranded template, a DNA polymerase enzyme, as well as deoxy- and dideoxy-nucleotides. The template was divided into four aliquots, having all four of the standard deoxynucleotides (dATP, dGTP, dCTP, and dTTP) and DNA polymerase. One of the four chain-terminating dideoxy-nucleotides (ddATP, ddGTP, ddCTP, or ddTTP) was added to each reaction to terminate the DNA strand during the chain elongation reaction. The method resulted in various length DNA fragments, which were heat denatured and size separated by denaturing polyacrylamide slab gel electrophoresis [62].

DNA sequence information has rapidly become essential for basic biological and applied clinical research and diagnosis, and it has significantly accelerated biomedical research and discovery by generating the complete genetic blueprints of human, many animal, plant, and microbial species. DNA-sequencing methods have recently evolved from the labor-intensive slab gel electrophoresis technique, through automated multicapillary electrophoresis systems using fluorophore labeling with multispectral imaging, to *next-generation* technologies of cyclic arrays, hybridization-based techniques, nanopore, and single molecule sequencing. Use of the current electrophoretic-based methodology for sequencing entire mammalian genomes, including standard molecular biology techniques for genomic sample preparation and the separation remains far too expensive to make genome sequencing affordable.

Using next-generation sequencing (NGS) approaches reduces DNA sequencing costs and increases throughput by employing massively parallel systems that can determine the sequence of huge numbers of different DNA strands simultaneously [63,64]. These systems allow millions of *reads* (contiguous regions of DNA sequence) to be gathered in a single experiment, in contrast to current multicapillary electrophoresis instruments, where the throughput is limited by the 96-capillary array (currently 700 high-quality bases/capillary/hour).

3.3.2 Next-generation sequencing

Next-generation sequencing technologies have further opened up the field of genomics, allowing new avenues of research to be pursued. In new generation sequencing, the aim is to develop higher-throughput sequencing methods and extracting/achieving more information from the genome. In addition, increasing sequencing speed and read length are also important to reach the ultimate goal of sequencing a genome for $1000 [36].

3.3.2.1   *Pyrosequencing*

The pyrosequencing process is performed in microfabricated high-density pL scale reactors achieving average read lengths of 500–800 bps with 1 million reads per run, resulting in 0.5–1 Gb information per run. Specific applications include *de novo* sequencing and resequencing of genomes, metagenomics, RNA analysis, and targeted sequencing of DNA regions of interest [65–68]. The complete sequencing workflow comprises four main steps, starting from purified DNA to analyzed results. These basic steps include generation of a single-stranded template DNA library, emulsion-based clonal amplification, data generation via sequencing-by-synthesis, and finally data analysis using advanced bioinformatics tools. The starting materials include genomic DNA, PCR products, BACs, and cDNA. Samples such as genomic DNA and BACs are fractionated into small, 300- to 800-base pair fragments. For smaller samples, such as small non-coding RNA or PCR amplicons, fragmentation is not required. A bottleneck of the technology is the identification of homopolymer regions, since where there are more than seven repeat units, the system is not capable of identifying the repeating bases because of the pyrosequencing chemistry. Using a series of standard molecular biology techniques, short adaptors—specific for both the 3′ and 5′ ends—are added to each fragment. The adaptors are used for purification, amplification, and sequencing steps. The single-stranded DNA library is immobilized onto specifically designed DNA capture beads. Each bead carries a unique single-stranded DNA library fragment. The bead-bound library is emulsified with amplification reagents in a water-in-oil mixture resulting in quasi-*microreactors* containing just one bead with one unique sample-library fragment. Each unique sample-library fragment is amplified within its own drop, excluding competing or contaminating sequences. Amplification of the entire fragment collection is done in parallel for each fragment, resulting in several million copies per bead. Subsequently, the emulsion PCR is broken by the addition of 2-butanol but the amplified fragments remain bound to their corresponding beads. The clonally amplified fragments are enriched and loaded onto the sequencing device. After the addition of sequencing enzymes, the fluidics subsystem of the sequencing instrument delivers the individual nucleotides in a fixed order across the hundreds of thousands of wells containing one bead each. The addition of nucleotides complementary to the template strand results in a chemiluminescent signal recorded by the charge-coupled device (CCD) camera within the instrument. For sequencing data analysis and assembly, a few bioinformatics tools are available supporting the following applications: *de novo* assembly

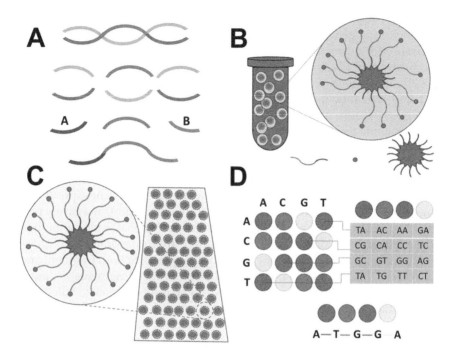

Figure 3.3 SOLiD technology. Fragmentation of a template DNA and specific adapter (A and B) ligation (A). Emulsion PCR in individual microreactors, which resulted in clonal amplification of each template molecule (B). The surface of the sequencing slide after loading the amplified template (C). The basis of color-coding (D). **(See color insert.)**

of up to 400 megabases; resequencing genomes of any size; and amplicon variant detection by comparison with a known reference sequence [36].

3.3.2.2 *The sequencing by oligo ligation/detection system*

This ligation-based method with a novel two-base color encoding system is significantly different from both traditional Sanger sequencing and the new high-throughput DNA sequencing technologies (Figure 3.3). Each bead contains PCR amplified copies of a single DNA molecule attached. Beads are then attached to the surface of a flow cell. Sequencing is accomplished then in a cyclic process though ligation of successive *probe* sequences to the beads. Each probe is ligated, read (by its fluorescence signal), and then appropriately cleared. Each ligation probe contains three degenerate bases (all possible 3 bp sequences synthesized combinatorically followed by a 2 bp read sequence, and then three fluorescently labeled universal bases [69]. Degenerate bases are used to provide a discrimination advantage during hybridization. Sequencing by Oligo

Ligation/Detection (SOLiD) system is very suitable for resequencing analysis including genomics, epigenetics, and metagenomics studies, and furthermore for performing gene expression and miRNA profiling studies. High-coverage RNA/transcriptome sequencing (digital gene expression) (DGE) allows gene expression-level and splice variant evaluation, as well as haplo-type ratio measurements in a qualitive and quantitative manner. However, because of the relatively short sequence reading, this approach is not suitable for *de novo* sequencing [22,70–73].

3.3.2.3   Sequencing by synthesis

Sequencing by synthesis (SBS) technology is one of the most successful and widely adopted next-generation sequencing approaches, providing the throughput of 200 Gb/10 days. SBS technology supports massively parallel sequencing using a proprietary reversible terminator-based method that enables detection of single bases as they are incorporated into growing DNA strands. SBS technology supports sequencing both fragment and paired-end libraries. Clonal clusters of DNA library fragments are automatically generated on planar, optically transparent substrates, called *flow cells*. A fluorescently labeled terminator is imaged as each dNTP is added and then cleaved to allow incorporation of the next base. Since all four reversible terminator-bound dNTPs are present during each sequencing cycle, natural competition minimizes incorporation bias. The end result is true base-by-base sequencing that enables the acquisition of accurate data for a broad range of applications. This technology is highly recommended for using genome and bacterial artificial chromosome (BAC) resequencing, ChIP sequencing, and furthermore, for gene expression profiling and small RNA identification [74,75].

3.4   Proteome analysis

3.4.1   Protein characterization techniques

Since its inception in the late 1960s, two-dimensional gel electrophoresis (2DE) represents one of the major analytical and preparative approaches to study complex protein mixtures [76,77]. Due to the limited sensitivity of the available visualization methods, even with the most sensitive silver or fluorescent staining techniques, 2DE can mostly show expression changes of medium- and high-abundant proteins. The separation power of 2DE is also limited by possible overlaps in mass and/or isoelectric point (pI) of the sample components. Other means of identification for protein spots of interest use approaches like Western blotting (Figure 3.4) and/or mass spectrometry (MS) after in-gel tryptic digestion [78,79].

It is important to note that 2DE separation and spot identification may be affected by posttranslational modifications and/or complex formation

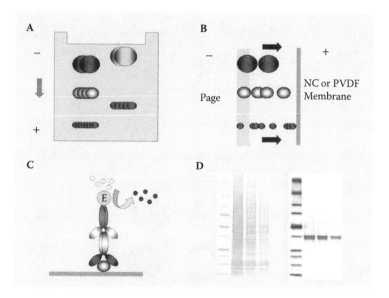

Figure 3.4 Western blotting technique. (A) Size separation of proteins by SDS-PAGE. (B) Transfer of the separated proteins onto nitrocellulose (NC) or poly-vinylidine fluoride (PVDF) membrane by electro blotting. (C) Visualization of proteins of interest by specific enzyme-conjugated antibody using immuno-staining. (D) Visualization of whole-protein extract by Coomassie Blue staining (left side), and immunostaining (right side), respectively.

with other biomolecules, while Western blotting is a simple and cost effec-tive semiquantitative method, only one protein can be characterized in one reaction, making it time consuming for larger-scale studies. MS is more effective but requires extra sample preparation steps (band excision, in-gel digestion, etc.) and expensive instrumentation with highly skilled personnel. Immunohistochemistry (IHC) is another option, which is capa-ble of localizing and identifying proteins of interest within a cell or tis-sue utilizing fluorescently labeled antibodies [80]. This approach can also be used to interrogate formalin-fixed, paraffin-embedded (FFPE) tissue samples even decades after their collection date. Although IHC is not able to quantify the exact amount of proteins, multispectral imaging allows examination of a number of different proteins in a single test. Combining IHC with rolling circle amplification (RCA) opens up a possibility for bet-ter quantification and the option to evaluate low-level, protein–protein interactions [81–83]. Rolling circle amplification is a method for single molecule visualization, for example, to detect one chromosome position or an individual cell surface protein, but it is also suitable for signal mul-tiplication (signal gain) in case of microarray (spots). Furthermore, IHC combined with tissue microarrays (TMA) (Figure 3.5) provides a new tool

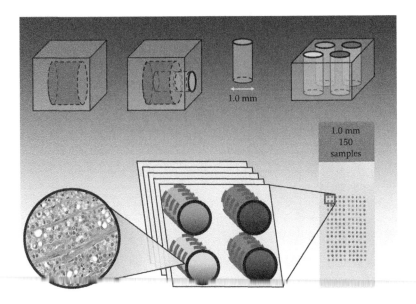

Figure 3.5 Schematic representation of tissue microarray (TMA) fabrication. The tissue cylinders are taken by a sampling needle and embedded into the acceptor block, followed by immunostaining and evaluation of staining intensity.

in molecular pathology and molecular diagnostics [84,85]. Tissue microarray technology, (TMT) or (TMA), is based on immunohistochemistry. It was named after the many tissue discs that were affixed to a microscope slide in a similar fashion as cDNA or protein microarrays, followed by interrogation with appropriate antibodies. In this way, the presence of a given protein can be simultaneously examined in a few dozen tissue samples. The tissue cylinders were made from formalin-fixed tissues embedded in paraffin. The diameter of the cylinders can range from 1 mm to almost 1 cm. Their size determines the number of samples that fit on a slide. The most common configurations are 4 × 6, 6 × 10, 8 × 12, and 10 × 24 [86,87]. The visualization and evaluation can be performed by conventional light microscope or using digitalized form by virtual microscope.

In quantitative proteomics studies, the most frequently employed method is enzyme-linked immunosorbent assay (ELISA) [88]. This sensitive and reliable, competition-based color reaction approach is widely used both in biomedical research and clinical diagnostics [89–91].

Table 3.2 summarizes the most popular proteome profiling techniques. Recent developments introduced capillary electrophoresis [92] not only for DNA analysis and sequencing, but also for high-resolution protein separation, based on their size (capillary electrophoresis-sodium dodecyl sulfate [CE-SDS]) and capillary isoelectric focusing (cIEF) point

[93,94]. Utilization of various affinity surfaces in conjunction with mass spectrometry is another recently emerging tool for proteome profiling and biomarker discovery. The functionalized surface chips are coupled with matrix-assisted laser desorption ionization time of flight mass spectrometry (MALDI-TOF), referred to as surface-enhanced laser desorption ionization (SELDI) [95]. Individual proteins or groups of proteins are bound to special affinity surfaces and embedded into an appropriate matrix for concomitant matrix-assisted laser desorption ionization. The ionized sample components are then analyzed by their mass to charge ratio in a time of flight (TOF) mass spectrometer [96]. This high-sensitivity method enables measurement of femto- or even attomol quantities of proteins. The dynamic range of the technique is limited to four orders of magnitude, which is not always satisfactory, especially in view of the dynamic range of the human plasma proteome that approximately covers 12 logs. On the other hand, proteome profile differences in the range of up to 300 kDa may be characteristic enough to define cellular states or diseases.

3.4.2   Protein microarrays

The first attempts to use protein microarrays simply employed ELISA technology. Later efforts utilized polyvinylidene fluoride (PVDF) or nitrocellulose (NC) membranes (filter array) with antibodies spotted and chemically fixed at specific locations [97]. During the interrogation of such membranes, various background problems arose mostly due to nonspecific protein binding, which was not diminished even by the application of stringency adjustment methods (varying salt concentration and/or temperature during the hybridization process). Blocking nonspecific binding sites prior to interrogation and/or using special surface treatments adequately addressed this problem [98]. The most commonly used protein microarrays employ immobilized antibodies interrogated by fluorescently labeled probes, or with known substrates to capture their antigens [97] (Figure 3.6). In reversed phase microarrays, known proteins are bound to a solid surface to capture antibodies from protein extracts, cell lysates, or other biological samples [13]. Microbead-based arrays, whole-cell arrays, and affinity microchips are working with similar principles [20,21,95].

3.4.2.1   Microarray detection methods

In the early days of protein microarray analysis, mostly radioisotope-based labeling was applied, especially for phosphorylated proteins [99]. Later, stable isotope-coded amino acids (SILAC) proved useful in *in vivo* incorporation to cell cultures [100]. Both methods were very sensitive, requiring only minute sample amounts to reveal protein expression differences [101]. The use of an enhanced chemiluminescence (ECL) technique using X-ray film exposure or a phosphor imager instrument was,

Table 3.2 Frequently Used Proteome Profiling Techniques

Test Method	Sample Amount	Cost	Number of Proteins	Advantages	Disadvantages	Comment
Gel electrophoresis	1–10 μg	Small	Few thousand	Costs, appropriate for screening	Labor-intensive, time-consuming	Multiple samples can be compared by using fluorescent labeling
Western blotting	40–100 μg	Modest	One	Protein identification, wide applicability	Poor reproducibility, time-consuming	Possible multiplexing
ELISA	10–1000 ng/96 well	Modest	One	Well established, automated sensitive, specific, small sample volume, wide applicability	Separate assay for each protein, different conditions for each plate, not suitable for small sample amounts	Multiplexing options
Immuno-histochemistry	Section from biopsies or surgical samples, swab	Small	One or maximum three per slide	Visualization of protein localization, easy sample preparation	Semiquantitative, special storage needed, limitation of retrospective studies	Freezing may cause poor morphology
TMA	Section from biopsies or surgical samples	Small	One per slide, 20–50/biopsies	Simultaneous analyis of multiple samples	Semiquantitative, not representative in case of small tissue spot	Easily automated, ideal for multiplex screening

Method	Sample amount	Sensitivity	Throughput	Advantages	Disadvantages	Comments
Planar protein chips (glass slide–based system)	20–100 µg crude extract	Reasonable	Few hundred proteins (Ab array) few thousand (RPA)	Small sample amounts, sensitive, relatively specific, multiplexing capabilities	Cross-reactivity, necessity of labeling	Antibody microarray, RPA, peptide array, aptamer array
Microbead-based array	10–500 ng	High	10–20 max 100	Multiplexing, small sample amounts, widely applicable in proteomics	Cross-reactivity, potential fluorescence interference	Currently used for cytokine-detection assays and autoantibody profiling
MALDI-MS	Under 1 µg or 10,000 cells	Very high	Few thousand	Small sample amounts, wide applicability, appropriate for screening	Equipment and labor-intensive, low reproducibility	Strong dependence on the sample preparation methods
SELDI-MS	Under 1 µg or 10,000 cells	Very high	Few thousand	Small sample amounts, appropriate for screening	Equipment and labor-intensive, low sensitivity	The mass spectrum obtained usually does not enable identification of the proteins
MELDI-MS	Under 1 µg or 10,000 cells	Very high	Few thousand	Small sample amounts, appropriate for screening	Equipment and labor-intensive	Some cases matrix free
SPR	10 pg–100 ng crude extract, 75 fg/ml–1 ng/ml protein	High	One or few	Small sample amounts, high sensitivity, no labeling requirement	Equipment and labor-intensive, inappropriate for screening	Kinetic measurements

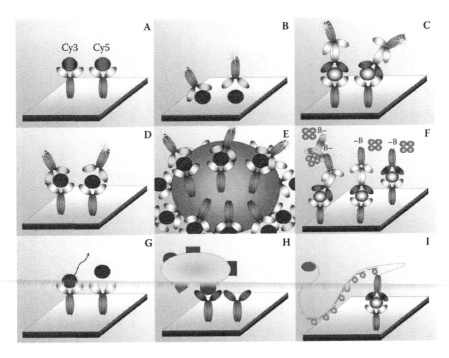

Figure 3.6 Summary diagram of protein microchip labeling/detection techniques and main areas, including: (A) Direct labeling. (B) Reversed-phase array (RPA). (C) Labeling with tertiary antibody. (D) Sandwich labeling. (E) Bead-based array. (F) Biotin-labeled antibody. (G) Label-free method for mass spectrometry. (H) Whole-cell array. (I) Rolling circle amplification (RCA). **(See color insert.)**

however, limited to one sample per test [78,79]. Advances in fluorescent labeling technology, such as the introduction of amine-reactive cyanine dyes (Cy3 and Cy5), changed the way protein microarrays are interrogated today [102]. The dye swap method is suitable for eliminating the disadvantages of dual labeling in antibody (AB) microarrays (Figure 3.7). Derivatization with these labeling methods provides identical fluorescent tagging efficiency that is advantageous in most comparative studies and also offers good quantification capabilities. With the use of different fluorophores (excitation and emission characteristics), many samples can be simultaneously analyzed in a single microarray. Detection is mostly done by fluorescence scanners or CCD cameras. The option of multispectral imaging can readily differentiate a larger number of fluorescent tags, provided that all emit at different wavelengths. With continuous improvement of the technology, protein microarrays today accommodate detection of several hundred or even several thousand micron-sized protein spots on a one square cm surface, making rapid and accurate experiments possible on a large scale [103].

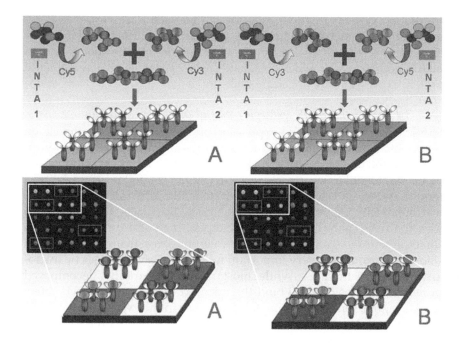

Figure 3.7 Dye swap labeling method for antibody microarrays. Upper panel: Antibody-containing slides are hybridized with fluorescently labeled sample proteins (A). In a parallel labeling reaction, the fluorescent dyes are swapped (B). Lower panel: Results: Green (A) or Red (B) spots: up-regulated proteins. Red (A) or Green (B) spots: down-regulated proteins. Yellow spots (A) and (B): no change in protein expression.

3.4.2.2   Sample preparation

Conventional glass slide-based protein arrays require relatively large sample amounts (10–200 µg crude protein sample) because of the labeling and cleaning process. A possible solution to this problem is the use of label-free methods or the application signal amplification methods [25]. While in the case of cell cultures and surgical specimens, where there is usually enough material available, it might represent a problem with some special clinical samples including FFPE samples. Other factors, such as problems with tissue homogeneity, may lead to false positive/negative results [104,105]. Analysis of circulating tumor cells (CTCs), amniotic fluid, and laser capture microdissected (LCM) samples are especially problematic [19,106–111] because of the low amount of sample (and consequently proteins) available. Unfortunately, for the time being, there is no similar amplification method for proteins as a polymerase chain reaction (PCR) for nucleic acids. Therefore, simultaneous protein expression and *in vitro* translation systems (without cells) should be used [112]. However, these mRNA-based methods are limited to providing only RNA-level information. Factors such as co- and posttranscriptional modifications, protein maturation processes, and mRNA half-life issues, which all may influence protein expression, are not addressed.

3.4.2.3   *False positives/negatives and protein degradation*

In microarray-based protein expression profiling, protein decomposition represents a significant issue. Proteins in cells are mostly present as short-lifetime effector molecules [32] and their degradation process is influenced by parameters like size, structure, composition, co- and posttranslational modifications, and so forth. Other factors like pH, salt concentration, hydrophobicity, or sample preparation-related artifacts, for example, in formalin-fixed, paraffin-embedded samples, may also cause changes in some protein recognition sites further emphasizing that conformational stability of proteins is very important for their reliable analysis by microarray-based techniques. While most truncated/partially decomposed protein molecules can still bind the microarray, they usually generate a weaker signal, thus, suggesting false decrease in the expression level [98]. In addition, false positives can be caused by the way the proteins immobilized to the microarray plate. Nonspecific binding is another factor that can cause inaccurate reading in protein microarray measurements. In antibody arrays, the antibody itself can be damaged during the fixation process, consequently showing decreased antigen recognition capability [113]. Finally, the presence of active proteolytic enzymes in the sample are also a potential cause of protein degradation.

3.4.2.4   *Protein microarray fabrication techniques*

The most crucial step during protein microarray fabrication is the fixation of proteins onto the array surface. The most commonly used wafer is glass, but silicon and gold surfaces [114,115], as well as special polymers, such as polycarbonate (PC), polymethyl methacrylate (PMMA), and polyethylene terephthalate (PET) [116–119] are also good alternatives. Figure 3.8 depicts the most frequently used capture molecule-immobilizing methods onto microarray surfaces: (A) adsorption (nitrocellulose, polyvinylidene fluoride, poly-L-lysine), (B) covalent binding (via various silane reagents), (C) affinity interactions (biotin/streptavidine, histidine-tag/nickel-nitrilotriacetic acid), and (D) diffusion-based (agarose, polyacrylamide, hydrogel) techniques. All of these immobilization/fixation approaches have their advantages and disadvantages. For example, adsorption is one of the simplest ways of immobilization, but it is relatively uncontrollable and also unstable. In addition, it may also lead to protein denaturation and cause steric hindrance [113]. Silanes are often used for protein immobilization in a covalent manner, but their binding

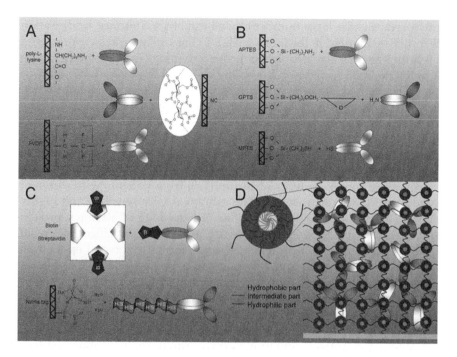

Figure 3.8 The most frequently used methods to immobilize capture molecules onto microarray surfaces. (A) Antibody adsorption by poly-L-lysine-, nitrocellulose-, or polyvinylidene fluoride-treated surface. (B) Covalent binding using various silane reagents of APTES (3-aminopropyl)triethoxysilane, GPTS (3-glycidoxypropyl)trimethoxysilane, and MPTS (3-mercaptopropyl)trimethoxysilane. (C) Affinity interactions by biotin/streptavidine or histidine-tag/nickel-nitrilotriacetic acid. (D) Diffusion-based (hydrogel) antibody-immobilization technique.

efficiency is very low [120]. Furthermore, they can also cause steric hindrance between the protein and the surface due to the length of spacer on the silane, reducing in this way the accessibility of the analyte to the affinity probe. One of the recent breakthroughs is the use of polyethylene terephthalate surface combined with polyglycidyl methacrylate (PGMA) photopolymer to covalently bind the proteins to immobilize probe molecules in high densities for high-performance protein microarrays [121].

3.4.3 *Most important protein microarray types*

3.4.3.1 *Glass microarray slides*

The most frequently used carriers in large-scale protein microarray studies based on protein–protein interactions (e.g., antibody/antigen recognition) are glass. The applicability of large surface areas enables interrogation of a great number of proteins at the same time. A particular advantage of glass microarrays is the large number of well-established surface modification techniques reported in the literature [113]. Glass microarrays are produced by printing small size droplets of protein solutions (e.g., antibodies) onto a pretreated glass surface, followed by chemical or physical immobilization. The method is, however, susceptible to possible cross-contamination (spray effect), and evaporation-mediated droplet segregation [122]. The two most frequently used carrier surfaces are chemically treated glass (for antibody attachment) and fixation of a special membrane onto the glass surface [98,113]. Hammerle and coworkers made glass-based protein arrays with 18 known auto-antigens in serial dilutions and hybridized with minimal amounts of serum [123] from patients with different types of autoimmune diseases. The assay was very sensitive and highly specific, proven by the detection of as little as 40 fg of a known protein standard with no apparent cross-reactivity. Others identified dozens of differentially expressed proteins between normal colorectal epithelium and adenocarcinoma by antibody microarray analysis [8]. These proteins gave a specific signature for colorectal cancer (CRC), different from other tumors, as well as a panel of novel markers along with potential targets for CRC. Antigen chips proved to be useful to distinguish different types of multiple sclerosis forms [124]. Based on a high-density myelin and inflammation-related, antigen-containing array, a unique auto-antibody pattern was determined, apparently suitable for identifying a certain multiple sclerosis subgroup. Approaches were also developed for enhanced protein microarray-based auto-antibody detection methods [125]. Cyanogen bromide (CNBr) digestion was used to reduce the complexity and size of the target proteins, thus increasing sensitivity and specificity. Reverse-phase microarray assays using phospho-specific antibodies are capable of directly measuring levels of phosphorylated protein isoforms. Samples from untreated colorectal cancer colectomies were laser capture microdissected to isolate epithelium and stroma from cancer as well as normal (i.e., uninvolved) mucosa. Lysates generated from these four tissue types were spotted onto reverse-phase protein microarrays and probed with a panel of antibodies. [126]. A multidimensional approach was developed for the identification of diagnostic antigens utilizing a combination of high-throughput antigen cloning and protein microarray-based serological detection of complex panels of antigens by exploiting the serum auto-antibody repertoire directed toward tumor-associated antigens in cancer patients [127]. The antigen-cloning technique is a method of choice to develop vaccines against cancer cells. The basis of this technology is that the antigens are able to directly stimulate the immune system helper T-cells. The cell surface receptors of the helper T-cells can recognize and bind tumor-related antigens. This process may help to eliminate tumor cells. An improved microarray approach was also reportedly coupled to microfluidics to detect allergen-specific immunoglobulin E (IgE) [128].

3.5 Other types of microarrays

3.5.1 MicroRNA analysis on microarray platforms

MicroRNA (miRNA) is a class of small single-stranded noncoding RNAs (19–30 nucleotides) that regulate gene expression by partial complementary base pairing to specific mRNAs. This structure facilitates the degradation of the target mRNA and in some cases physically inhibits protein translation. MicroRNAs have been found in animals, plants, and viruses. As of today, over 1500 miRNAs have been identified in humans according to the Sanger miRBase v18 [129,130]. MicroRNA genes are transcribed as pre-miRNAs, which are then processed to shorter hairpin-shaped pre-miRNAs (~70–90 nucleotides) and cleaved to form the mature single-stranded miRNAs (~22 nucleotides). As part of a multiprotein RNA-induced silencing complex (RISC), miRNAs repress translation by forming imperfect base pairing with the 3′ untranslated region of the target messenger RNA (mRNAs). Bioinformatics and cloning studies have estimated that miRNAs may regulate the expression of up to 30% of all human genes [131–136].

One of the most commonly used multiplex identification and/ or screening methods for miRNAs is microarray technology based. Complementary oligonucleotides are synthesized on the array surface to capture the labeled miRNA components from cell lysates. Enriched low molecular weight (LMW) RNA and total RNA are both suitable for the hybridization method because of the relatively high abundance of miRNAs in the cells. Microarray technology is a powerful tool for studying the role of miRNAs in normal and disease states. In addition, miRNA studies facilitate the discovery of biomarkers and disease signatures, in some cases even more reliably than that of protein or mRNA profiling methods. MiRNA arrays may also be able to detect small nucleolar ribonucleic acids (snoRNAs) and small Cajal body-specific ribonucleic acids (scaRNAs), which are short, nontranslated RNAs reportedly playing a role in biogenesis and the modification of ribosomal RNAs. SnoRNAs have also been implicated in the regulation of alternative splicing [137].

Wang et al. have developed a sensitive, accurate, and multiplexed microRNA profiling assay, based on a highly efficient labeling method and novel microarray probe design [138]. The use of specially designed probes (that can establish the loop region based on self-complementer sequence regions, which are suitable for distinguishing the mature miRNA from other RNAs, for example, mRNA, pre- and pri-miRNA) in most cases enabled highly specific detection of closely related mature miRNAs. Recent advances in contemporary inkjet technologies enabled direct synthesis of 40- to 60-mer oligonucleotide probes on the array surface, resulting in high-fidelity, self-loop forming probes, which provide both sequence and size discrimination with directly labeled RNA samples. Bead-based arrays are also introduced for miRNA detection and quantification [139]. This reportedly faster method provided higher reproducibility than competing solid-phase kinetics limited planar arrays [140]. However, distinguishing and quantitating miRNA isoforms from the hybridization-based platforms are not always suitable because of the large homologous sequence variants and very similar RNA molecules; therefore, individual assays must be used to validate the screening methods. The locked nucleic acid (LNA) oligo-based Northern hybridization approach and the more recently introduced stem loop-based RT-PCR are both capable of validating microarray results [141–143].

3.5.2 Affinity surface arrays

Surface-enhanced laser desorption ionization (SELDI) chips [21] were introduced in the late 1990s (Figure 3.9). These devices incubate the sample of interest with an affinity surface and analyze the bound proteins by time of flight mass spectrometry (TOF-MS). In addition to proteomics profiling, the approach was also capable of analyzing protein–protein interactions. With the use of special chromatographic surfaces, extent and profile of phosphorylation, peptide fingerprinting, and epitope mapping

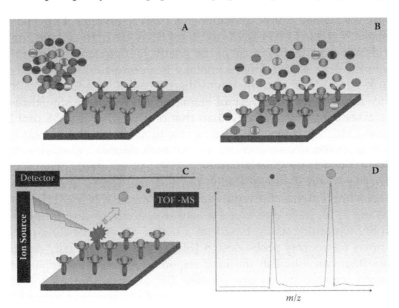

Figure 3.9 Schematics of the use of affinity protein microchips. Upper panel: Sample introduction (A) and affinity binding (B) of proteins of interest. Lower panel: Laser desorption/ionization (C) and MS interrogation (D).

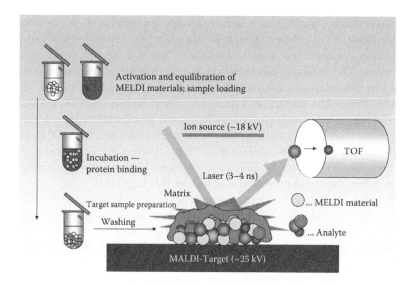

Figure 3.10 Material-enhanced laser desorption/ionization (MELDI)-based biomarker discovery. (Courtesy of C. Huck and M. Rainer.)

were also reported [144,145]. The SELDI-TOF-MS method was compared to 2D electrophoresis images in pharynx normal epithelium and squamous tumor tissues for biomarker discovery [146] and also used for quantitative evaluation of transthyretin (TTR) variants, both in tissue amyloid deposits and body fluids [147].

Another novel approach to utilizing affinity surfaces is referred to as the *material-enhanced laser desorption/ionization* (MELDI) technique [95] (Figure 3.10). This method was developed with the use of specially derivatized chromatographic materials (cellulose, silica, poly [glycidyl methacrylate/divinylbenzene] particles, and diamond powder) for fast and direct protein profiling to evaluate normal and pathological human serum samples.

3.5.3   Biosensor chips and microbeads

Surface plasmon resonance (SPR) is one of the most sophisticated methods these days for real-time detection and quantification of biomolecular interactions in a nondestructive fashion and without any labeling requirement [148]. SPR is based on the detection of binding pattern changes on a specific sensor surface caused by the interaction between the solute molecules in solution and the immobilized affinity components on the surface. Glass surfaces are coated by a matrix (e.g., dextrin) along with a conductive gold film layer, where the capture molecules are immobilized. A continuous analyte flow is driven through the sensor chip providing the best possible interaction between the analyte molecules and the immobilized capture molecules. The interrogation surface is regenerated after each measurement. For downstream analysis, the selectively captured molecules can be recovered from the surface and analyzed, for example, by mass spectrometry [149]. This approach proved to be useful in biomolecular interaction pattern analysis including drug candidates, enzyme inhibitors, DNA binding proteins, posttranslational modifications, disease markers, and specific peptides [150–153]. Recently, SPR was combined with rolling circle amplification using nanogold-modified oligonucleotide tags [154].

In microbead arrays, capture molecules attached onto microbead surfaces of, for example, polystyrene microspheres, act as protein microarrays and thus, can be used as such in protein–protein, nucleic acid–protein, and nucleic acid–nucleic acid interaction studies using flow cytometry-based interrogation. Multiplexing is possible by the utilization of sized [20] or color-coded microbeads [155].

3.5.4   Aptamers and aptamer arrays

Aptamers were first described in 1990 [156–158] as a special type of synthetic oligonucleotide molecule (ribonucleic acids or deoxyribonucleic acids) with a unique three-dimensional structure that specifically binds target molecules. Aptamers are typically tens to hundreds of bases long. Through a process termed *Systemic Evolution of Ligands by Exponential Enrichment* (SELEX) [159], complex libraries of up to 10^{15} nucleic acid oligomers can be screened for their ability to interact with a given target molecule [160]. Similar to antibodies, aptamers bind their target with high specificity and high affinity. While a variety of analytical applications of aptamers have already been demonstrated, their commercial development has not matured yet. Aptamer technology is an emerging chemical and biological recognition approach that offers several advantages over traditional bioreceptors, such as they can be formulated to bind specific targets, including small molecules, proteins, nucleic acids, metabolites, viruses, toxins, cells, and spores [161]. Although the mechanism of aptamer binding is similar to that of antibody–antigen interactions, unlike antibodies, aptamers are amenable to mass production without the need of immunization [162]. Other distinct advantages of aptamers over antibody-based techniques include greater stability for field use, and the ease of possible chemical modifications [163]. Several unique properties of aptamers, including high-binding specificity, low immunogenicity, structural stability, and ease of synthesis have made them promising agents for directed therapy against cancer targets. Aptamers with antineoplastic activity against extracellular, cell membrane, and intracellular targets have all been developed and incorporated into novel constructs as tools for delivering therapeutic agents. They can act as antibody substitutes in standard applications such as ELISA, Western blot, a fluorescence-activated cell sorter (FACS), and antibody microarrays [164].

3.5.5 Lectin arrays

Carbohydrates are very important classes of biomolecules in life, and glycosylation is apparently one of the most abundant co- and posttranslational modifications of proteins. Glycosylation has a structural role as well as diverse functional roles in many specific biological functions, including cancer development, viral and bacterial infections, and autoimmunity [165]. The increasing amount of research and the consequent rise in the commercialization activity in the glycomics field indicates that glycoscience is an emerging field in the arena of modern biotechnology. In less than a decade, glycomics has also taken center stage in novel approaches to healthcare. The aim of glycobiology is to study how carbohydrates play a vital role in living systems. However, glycobiology research is limited by the lack of rapid analysis options of the glycan moiety in glycoproteins. Currently used chromatography-, electrophoresis-, and mass spectrometry-based methods for glycoanalysis are tedious, time consuming, and typically require high levels of expertise. The advent of lectin arrays enabled rapid glyco-type analysis. The method is based on binding intact glycoproteins to surface arrayed lectins, resulting in a characteristic fingerprint that is highly perceptive to changes in the glycan composition on the proteins. The large number of available lectins, each with its specific structural recognition feature, ensures appropriate detection of general changes in glycosylation patterns [166–168].

For fabrication of carbohydrate microarrays with monosaccharides, oligosaccharides, and polysaccharides, carbohydrates can be substrate specifically designed. This substrate specificity allows the fabrication of carbohydrate microarrays with a wide range of structurally diverse carbohydrate probes (including linkage and positional isomer specificity) without any chemical modification requirement [169,170].

Water soluble, stable, and biologically active glyco-nanoparticles have been developed for *in vivo* and *in vitro* diagnostic screening and therapeutic techniques [171–173]. The approach is based on the conjugation of carbohydrates onto nanoparticle surfaces such as fluorescent semiconductor quantum dots (QDs), gold nanoparticles, and magnetic nanoparticles. These glyco-nanoparticles offer a large number of possible applications in the biological and biomedical fields, for example, glyco-QDs present the carbohydrate antigens in a three-dimensional and polyvalent array, conferring biological specificity [174,175]. Among the numerous methods for preparing carbohydrate microarrays described in the literature are nitrocellulose-coated slides for noncovalent immobilization of microbial polysaccharides and neoglycolipid-modified oligosaccharides, polystyrene microtiter plates for lipid-bearing carbohydrates, self-assembled monolayers modified by Diels–Alder-mediated coupling of cyclopentadiene-derivatized saccharides, thiol-derivatized glass slides modified with maleimide-functionalized oligosaccharides, and thiol-functionalized carbohydrates immobilized on maleimide-derived gold slides, just to list the more important ones [169,176–178].

In addition to lectins, phage display is another promising tool for glycomics analysis [179,180]. Phage display is an interaction-based approach for selecting peptides, proteins, or antibodies with specific binding properties. Sequence modification of bacteriophage DNA can result in a required protein expression on the phage particle/capsid surface. This method is suitable for studying protein–protein interactions from a large number of variants.

3.6   Future perspectives

Microarray technology provides high-throughput and high-performance tools to monitor and determine global expression levels at the genome, proteome, glycome, and metabolome levels and to investigate interactions between biologically important molecules, making this technology especially useful in chemical genomics and chemical proteomics studies. Therefore, microchips and microarrays are of high interest and getting closer and closer to being available for routine biomedical and diagnostic applications provided that the ongoing technological developments are successful in improving sensitivity and specificity. However, a real breakthrough in their use is not expected until costs are reduced by at least a factor of 10 or even more. At the present time, the majority of complex genetic diseases are not well defined and protein microchips hold the promise of revealing the missing information for better understanding in areas like signaling pathways and apoptosis, as well as shedding light on many effector and pleiotropic functions. The use of state of the art bioinformatics tools enables full utilization of microarray results among RNA expression chip data, DNA sequence information, and protein microchips in a comparative manner. Bioinformatics-based cross-validation of cellular models with signaling pathways and over/under expression at the transcriptome and proteome levels, interrogated by microarray techniques will help in understanding and identifying key genes, gene modifications, and crucial nodal points as possible drug targets when chemical genomics and chemical proteomics approaches are used.

References

1. Gold, L., D. Ayers, J. Bertino, C. Bock, A. Bock, E. N. Brody, J. Carter, A. B. Dalby, B. E. Eaton, T. Fitzwater, D. Flather, et al. 2010. Aptamer-based multiplexed proteomic technology for biomarker discovery. *PLoS One* 5:15004.
2. Lujambio, A., A. Portela, J. Liz, S. A. Melo, S. Rossi, R. Spizzo, C. M. Croce, G. A. Calin, and M. Esteller. 2010. CpG island hypermethylation-associated silencing of non-coding RNAs transcribed from ultraconserved regions in human cancer. *Oncogene* 29:6390–401.
3. Oyelaran, O. and J. C. Gildersleeve. 2010. Evaluation of human antibody responses to keyhole limpet hemocyanin on a carbohydrate microarray. *Proteom. Clin. Appl.* 4:285–94.
4. Riveros, C., D. Mellor, K. S. Gandhi, F. C. McKay, M. B. Cox, R. Berretta, S. Y. Vaezpour, M. Inostroza-Ponta, S. A. Broadley, et al. 2011. A transcription factor map as revealed by a genome-wide gene expression analysis of whole-blood mRNA transcriptome in multiple sclerosis. *PLoS One* 5:14176.
5. Meany, D. L., L. Hackler, H. Zhang, and D. Chan. 2011. Tyramide signal amplification for antibody-overlay lectin microarray: A strategy to improve the sensitivity of targeted glycan profiling. *J. Proteome Res.* 10(3):1425–31.
6. Lukk, M., M. Kapushesky, J. Nikkila, H. Parkinson, A. Goncalves, W. Huber, E. Ukkonen, and A. Brazma. 2011. A global map of human gene expression. *Nat. Biotechnol.* 28:322–4.
7. Wiese, R., Y. Belosludtsev, T. Powdrill, P. Thompson, and M. Hogan. 2001. Simultaneous multianalyte ELISA performed on a microarray platform. *Clin. Chem.* 47:1451–7.
8. Madoz-Gurpide, J., M. Canamero, L. Sanchez, J. Solano, P. Alfonso, and J. I. Casal. 2007. A proteomics analysis of cell signaling alterations in colorectal cancer. *Mol. Cell Proteomics* 6:2150–64.
9. Spisak, S., B. Galamb, B. Wichmann, F. Sipos, O. Galamb, N. Solymosi, B. Nemes, Z. Tulassay, and B. Molnar. 2009. Tissue microarray (TMA) validated progression markers in colorectal cancer using antibody microarrays. *Orv. Hetilap.* 150: 1607–13.
10. Walter, G., K. Bussow, A. Lueking, and J. Glokler. 2002. High-throughput protein arrays: Prospects for molecular diagnostics. *Trends Mol. Med.* 8:250–3.
11. Schweitzer, B., S. Wiltshire, J. Lambert, S. O'Malley, K. Kukanskis, Z. Zhu, S. F. Kingsmore, P. M. Lizardi, and D. C. Ward. 2000. Inaugural article: Immunoassays with rolling circle DNA amplification: A versatile platform for ultrasensitive antigen detection. *P. Natl. Acad. Sci. USA* 97:10113–9.
12. Liu, Q., Q. Chen, J. Wang, Y. Zhang, Y. Zhou, C. Lin, W. He, F. He, and D. Xu. 2010. A miniaturized sandwich immunoassay platform for the detection of protein–protein interactions. *BMC Biotechnol.* 10:78.
13. Zha, H., M. Raffeld, L. Charboneau, S. Pittaluga, L. W. Kwak, E. Petricoin, 3rd, L. A. Liotta, and E. S. Jaffe. 2004. Similarities of prosurvival signals in Bcl-2-positive and Bcl-2-negative follicular lymphomas identified by reverse phase protein microarray. *Lab. Invest.* 84:235–44.
14. Espina, V., J. Wulfkuhle, V. S. Calvert, L. A. Liotta, and E. F. Petricoin, 3rd. 2008. Reverse phase protein microarrays for theranostics and patient-tailored therapy. *Methods Mol. Biol.* 441:113–28.

15. Sheehan, K. M., V. S. Calvert, E. W. Kay, Y. Lu, D. Fishman, V. Espina, J. Aquino, R. Speer, R. Araujo, G. B. Mills, L. A. Liotta, E. F. Petricoin, 3rd, and J. D. Wulfkuhle. 2005. Use of reverse phase protein microarrays and reference standard development for molecular network analysis of metastatic ovarian carcinoma. *Mol. Cell Proteomics* 4:346–55.

16. Popova, T. G., M. J. Turell, V. Espina, K. Kehn-Hall, J. Kidd, A. Narayanan, L. Liotta, E. F. Petricoin, 3rd, F. Kashanchi, C. Bailey, and S. G. Popov. 2010. Reverse-phase phosphoproteome analysis of signaling pathways induced by Rift valley fever virus in human small airway epithelial cells. *PLoS One* 5:13805.

17. Wilson, B., L. A. Liotta, and E. Petricoin, 3rd. 2010. Monitoring proteins and protein networks using reverse phase protein arrays. *Dis. Markers* 28:225–32.

18. de Wildt, R. M., C. R. Mundy, B. D. Gorick, and I. M. Tomlinson. 2000. Antibody arrays for high-throughput screening of antibody-antigen interactions. *Nat. Biotechnol.* 18:989–94.

19. Nagrath, S., L. V. Sequist, S. Maheswaran, D. W. Bell, D. Irimia, L. Ulkus, M. R. Smith, E. L. Kwak, S. Digumarthy, et al. 2007. Isolation of rare circulating tumour cells in cancer patients by microchip technology. *Nature* 450:1235–9.

20. Morgan, E., R. Varro, H. Sepulveda, J. A. Ember, J. Apgar, J. Wilson, L. Lowe, R. Chen, L. Shivraj, A. Agadir, R. Campos, D. Ernst, and A. Gaur. 2004. Cytometric bead array: A multiplexed assay platform with applications in various areas of biology. *Clin. Immunol.* 110:252–66.

21. Kuwata, H., T. T. Yip, C. L. Yip, M. Tomita, and T. W. Hutchens. 1998. Bactericidal domain of lactoferrin: Detection, quantitation, and characterization of lactoferricin in serum by SELDI affinity mass spectrometry. *Biochem. Biophys. Res. Commun.* 245:764–73.

22. Khan, I. H., S. Mendoza, P. Rhyne, M. Ziman, J. Tuscano, D. Eisinger, H. J. Kung, and P. A. Luciw. 2006. Multiplex analysis of intracellular signaling pathways in lymphoid cells by microbead suspension arrays. *Mol. Cell Proteomics* 5:758–68.

23. Gu, Y., A. Zeleniuch-Jacquotte, F. Linkov, K. L. Koenig, M. Liu, L. Velikokhatnaya, R. E. Shore, A. Marrangoni, P. Toniolo, A. E. Lokshin, and A. A. Arslan. 2009. Reproducibility of serum cytokines and growth factors. *Cytokine* 45:44–9.

24. Schwenk, J. M., U. Igel, B. S. Kato, G. Nicholson, F. Karpe, M. Uhlen, and P. Nilsson. 2010. Comparative protein profiling of serum and plasma using an antibody suspension bead array approach. *Proteomics* 10:532–40.

25. Spisak, S., Z. Tulassay, B. Molnar, and A. Guttman. 2007. Protein microchips in biomedicine and biomarker discovery. *Electrophoresis* 28:4261–73.

26. Fodor, S. P., R. P. Rava, X. C. Huang, A. C. Pease, C. P. Holmes, and C. L. Adams. 1993. Multiplexed biochemical assays with biological chips. *Nature* 364:555–6.

27. Davis, S. and P. S. Meltzer. 2007. GEOquery: A bridge between the Gene Expression Omnibus (GEO) and BioConductor. *Bioinformatics* 23:1846–1847.

28. Galamb, O., F. Sipos, N. Solymosi, S. Spisak, T. Krenacs, K. Toth, Z. Tulassay, and B. Molnar. 2008. Diagnostic mRNA expression patterns of inflamed, benign, and malignant colorectal biopsy specimen and their correlation with peripheral blood results. *Cancer Epidemiol. Biomarkers Prev.* 17:2835–45.

29. Zhang, L., W. Zhou, V. E. Velculescu, S. E. Kern, R. H. Hruban, S. R. Hamilton, B. Vogelstein, and K. W. Kinzler. 1997. Gene expression profiles in normal and cancer cells. *Science* 276:1268–72.

30. Helm, M. 2006. Post-transcriptional nucleotide modification and alternative folding of RNA. *Nucleic Acids Res.* 34:721–33.
31. Galamb, O., F. Sipos, S. Spisak, B. Galamb, T. Krenacs, G. Valcz, Z. Tulassay, and B. Molnar. 2009. Potential biomarkers of colorectal adenoma-dysplasia-carcinoma progression: mRNA expression profiling and *in situ* protein detection on TMAs reveal 15 sequentially upregulated and 2 downregulated genes. *Cell Oncol.* 31:19–29.
32. Aufderheide, A. C., W. Salo, M. Madden, J. Streitz, J. Buikstra, F. Guhl, B. Arriaza, C. Renier, L. E. Wittmers, Jr., G. Fornaciari, and M. Allison. 2004. A 9,000-year record of Chagas' disease. *P. Natl. Acad. Sci. USA* 101:2034–9.
33. Spisak, S., B. Galamb, F. Sipos, O. Galamb, B. Wichmann, N. Solymosi, B. Nemes, J. Molnar, Z. Tulassay, and B. Molnar. 2010. Applicability of antibody and mRNA expression microarrays for identifying diagnostic and progression markers of early and late stage colorectal cancer. *Dis. Markers* 28:1–14.
34. Huang, R. P., R. Huang, Y. Fan, and Y. Lin. 2001. Simultaneous detection of multiple cytokines from conditioned media and patient's sera by an antibody-based protein array system. *Anal. Biochem.* 294:55–62.
35. Venter, J. C., M. D. Adams, E. W. Myers, P. W. Li, R. J. Mural, G. G. Sutton, H. O. Smith, M. Yandell, C. A. Evans, R. A. Holt, et al. 2001. The sequence of the human genome. *Science* 291: 1304–51.
36. Hert, D. G., C. P. Fredlake, and A. E. Barron. 2008. Advantages and limitations of next-generation sequencing technologies: A comparison of electrophoresis and non-electrophoresis methods. *Electrophoresis* 29:4618–26.
37. Lander, E. S., L. M. Linton, B. Birren, C. Nusbaum, M. C. Zody, J. Baldwin, K. Devon, K. Dewar, M. Doyle, W. FitzHugh, R. Funke, D. Gage, et al. 2001. Initial sequencing and analysis of the human genome. *Nature* 409:860–921.
38. Istrail, S., G. G. Sutton, L. Florea, A. L. Halpern, C. M. Mobarry, R. Lippert, B. Walenz, H. Shatkay, I. Dew, J. R. Miller, M. J. Flanigan, et al. 2004. Whole-genome shotgun assembly and comparison of human genome assemblies. *Proc. Natl. Acad. Sci. USA* 101:1916–21.
39. Pascal, L. E., L. D. True, D. S. Campbell, E. W. Deutsch, M. Risk, I. M. Coleman, L. J. Eichner, P. S. Nelson, and A. Y. Liu. 2008. Correlation of mRNA and protein levels: Cell type-specific gene expression of cluster designation antigens in the prostate. *BMC Genomics* 9:246.
40. Galamb, O., B. Molnar, and Z. Tulassay. 2003. [DNA chips for gene expression analysis and their application in diagnostics]. *Orv. Hetilap.* 144:21–7.
41. LeProust, E. M., B. J. Peck, K. Spirin, H. B. McCuen, B. Moore, E. Namsaraev, and M. H. Caruthers. 2010. Synthesis of high-quality libraries of long (150mer) oligonucleotides by a novel depurination controlled process. *Nucleic Acids Res.* 38: 2522–40.
42. Nuwaysir, E. F., W. Huang, T. J. Albert, J. Singh, K. Nuwaysir, A. Pitas, T. Richmond, T. Gorski, J. P. Berg, J. Ballin, M. McCormick, et al. 2002. Gene expression analysis using oligonucleotide arrays produced by maskless photolithography. *Genome Res.* 12:1749–55.
43. Barone, A. D., J. E. Beecher, P. A. Bury, C. Chen, T. Doede, J. A. Fidanza, and G. H. McGall. 2001. Photolithographic synthesis of high-density oligonucleotide probe arrays. *Nucleos Nucleot. Nucl.* 20:525–31.
44. McGall, G. H. and J. A. Fidanza, 2001. Photolithographic synthesis of high-density oligonucleotide arrays. *Methods Mol. Biol.* 170:71–101.

45. Singh-Gasson, S., R. D. Green, Y. Yue, C. Nelson, F. Blattner, M. R. Sussman, and F. Cerrina. 1999. Maskless fabrication of light-directed oligonucleotide microarrays using a digital micromirror array. *Nat. Biotechnol.* 17:974–8.

46. Lipshutz, R. J., D. Morris, M. Chee, E. Hubbell, M. J. Kozal, N. Shah, N. Shen, R. Yang, and S. P. Fodor. 1995. Using oligonucleotide probe arrays to access genetic diversity. *Biotechniques* 19:442–7.

47. Chee, M., R. Yang, E. Hubbell, A. Berno, X. C. Huang, D. Stern, J. Winkler, D. J. Lockhart, M. S. Morris, and S. P. Fodor. 1996. Accessing genetic information with high-density DNA arrays. *Science* 274:610–4.

48. Rollins, B., M. V. Martin, P. A. Sequeira, E. A. Moon, L. Z. Morgan, S. J. Watson, A. Schatzberg, H. Akil, R. M. Myers, E. G. Jones, D. C. Wallace, W. E. Bunney, and M. P. Vawter. 2009. Mitochondrial variants in schizophrenia, bipolar disorder, and major depressive disorder. *PLoS One* 4:4913.

49. Corless, C. E., E. Kaczmarski, R. Borrow, and M. Guiver. 2008. Molecular characterization of Neisseria meningitidis isolates using a resequencing DNA microarray. *J. Mol. Diagn.* 10:265–71.

50. Fokstuen, S., R. Lyle, A. Munoz, C. Gehrig, R. Lerch, A. Perrot, K. J. Osterziel, C. Geier, M. Beghetti, F. Mach, J. Sztajzel, U. Sigwart, S. E. Antonarakis, and J. L. Blouin. 2008. A DNA resequencing array for pathogenic mutation detection in hypertrophic cardiomyopathy. *Hum. Mutat.* 29:879–85.

51. Lebet, T., R. Chiles, A. P. Hsu, E. S. Mansfield, J. A. Warrington, and J. M. Puck. 2008. Mutations causing severe combined immunodeficiency: Detection with a custom resequencing microarray. *Genet. Med.* 10:575–85.

52. Waldmuller, S., M. Muller, K. Rackebrandt, P. Binner, S. Poths, M. Bonin, and T. Scheffold. 2008. Array-based resequencing assay for mutations causing hypertrophic cardiomyopathy. *Clin. Chem.* 54:682–7.

53. Pani, A. M., H. H. Hobart, C. A. Morris, C. B. Mervis, P. Bray-Ward, K. W. Kimberley, C. M. Rios, R. C. Clark, M. D. Gulbronson, G. C. Gowans, and R. G. Gregg. 2011. Genome rearrangements detected by SNP microarrays in individuals with intellectual disability referred with possible Williams syndrome. *PLoS One* 5:12349.

54. Sayagues, J. M., C. Fontanillo, M. Abad Mdel, M. Gonzalez-Gonzalez, M. E. Sarasquete, C. Chillon Mdel, E. Garcia, O. Bengoechea, E. Fonseca, M. Gonzalez-Diaz, J. De las Rivas, L. Munoz-Bellvis, and A. Orfao. 2011. Mapping of genetic abnormalities of primary tumours from metastatic CRC by high-resolution SNP arrays. *PLoS One* 5:13752.

55. Ronald, A., L. M. Butcher, S. Docherty, O. S. Davis, L. C. Schalkwyk, I. W. Craig, and R. Plomin. 2010. A genome-wide association study of social and non-social autistic-like traits in the general population using pooled DNA, 500 K SNP microarrays and both community and diagnosed autism replication samples. *Behav. Genet.* 40:31–45.

56. Yan, M., Z. Zhang, J. R. Brady, S. Schilbach, W. J. Fairbrother, and V. M. Dixit. 2002. Identification of a novel death domain-containing adaptor molecule for ectodysplasin-A receptor that is mutated in crinkled mice. *Curr. Biol.* 12:409–13.

57. Szponar, A., D. Zubakov, J. Pawlak, A. Jauch, and G. Kovacs. 2009. Three genetic developmental stages of papillary renal cell tumors: Duplication of chromosome 1q marks fatal progression. *Int. J. Cancer* 124:2071–6.

58. Yagi, K., K. Akagi, H. Hayashi, G. Nagae, S. Tsuji, T. Isagawa, Y. Midorikawa, Y. Nishimura, H. Sakamoto, Y. Seto, H. Aburatani, and A. Kaneda. 2010. Three DNA methylation epigenotypes in human colorectal cancer. *Clin. Cancer Res.* 16:21–33.

59. Van der Auwera, I., W. Yu, L. Suo, L. van Neste, P. van Dam, E. A. van Marck, P. Pauwels, P. B. Vermeulen, L. Y. Dirix, and S. J. van Laere. 2010. Array-based DNA methylation profiling for breast cancer subtype discrimination. *PLoS One* 5(9):e12616.

60. Taskesen, E., R. Beekman, J. de Ridder, B. J. Wouters, J. K. Peeters, I. P. Touw, M. J. Reinders, and R. Delwel. 2010. HAT: Hypergeometric analysis of tiling-arrays with application to promoter-GeneChip data. *BMC Bioinformatics* 11:275.

61. Unver, T., M. Bakar, R. C. Shearman, and H. Budak. 2010. Genome-wide profiling and analysis of *Festuca arundinacea* miRNAs and transcriptomes in response to foliar glyphosate application. *Mol. Genet. Genomics* 283:397–413.

62. Karger, B. L. and A. Guttman. 2009. DNA sequencing by CE. *Electrophoresis* 30(Suppl. 1):S196–S202.

63. Hutchison, C. A., 3rd. 2007. DNA sequencing: Bench to bedside and beyond. *Nucleic Acids Res.* 35:6227–37.

64. Rogers, Y. H. and J. C. Venter. 2005. Genomics: Massively parallel sequencing. *Nature* 437:326–7.

65. Parameswaran, P., R. Jalili, L. Tao, S. Shokralla, B. Gharizadeh, M. Ronaghi, and A. Z. Fire. 2007. A pyrosequencing-tailored nucleotide barcode design unveils opportunities for large-scale sample multiplexing. *Nucleic Acids Res.* 35:130.

66. Zagrobelny, M., K. Scheibye-Alsing, N. B. Jensen, B. L. Moller, J. Gorodkin, and S. Bak. 2009. 454 pyrosequencing based transcriptome analysis of *Zygaena filipendulae* with focus on genes involved in biosynthesis of cyanogenic glucosides. *BMC Genomics* 10:574.

67. Peng, Y., L. L. Abercrombie, J. S. Yuan, C. W. Riggins, R. D. Sammons, P. J. Tranel, and C. N. Stewart, Jr. 2010. Characterization of the horseweed (*Conyza canadensis*) transcriptome using GS-FLX 454 pyrosequencing and its application for expression analysis of candidate non-target herbicide resistance genes. *Pest Manag. Sci.* 66:1053–62.

68. Petrosino, J. F., S. Highlander, R. A. Luna, R. A. Gibbs, and J. Versalovic. 2009. Metagenomic pyrosequencing and microbial identification. *Clin. Chem.* 55:856–66.

69. Harismendy, O., P. C. Ng, R. L. Strausberg, X. Wang, T. B. Stockwell, K. Y. Beeson, N. J. Schork, S. S. Murray, E. J. Topol, S. Levy, and K. A. Frazer. 2009. Evaluation of next generation sequencing platforms for population targeted sequencing studies. *Genome Biol.* 10:R32.

70. Lao, K. Q., F. Tang, C. Barbacioru, Y. Wang, E. Nordman, C. Lee, N. Xu, X. Wang, B. Tuch, J. Bodeau, A. Siddiqui, and M. A. Surani. 2009. mRNA-sequencing whole transcriptome analysis of a single cell on the SOLiD system. *J. Biomol. Technol.* 20:266

71. Fanelli, M., S. Amatori, I. Barozzi, M. Soncini, R. Dal Zuffo, G. Bucci, M. Capra, M. Quarto, G. I. Dellino, C. Mercurio, M. Alcalay, G. Viale, P. G. Pelicci, and S. Minucci. 2010. Pathology tissue-chromatin immunoprecipitation, coupled with high-throughput sequencing, allows the epigenetic profiling of patient samples. *P. Natl. Acad. Sci. USA* 107:21535–40.

72. Hoffman, B. G. and S. J. Jones. 2009. Genome-wide identification of DNA-protein interactions using chromatin immunoprecipitation coupled with flow cell sequencing. *J. Endocrinol.* 201:1–13.

73. Kaufmann, K., J. M. Muino, M. Osteras, L. Farinelli, P. Krajewski, and G. C. Angenent. 2010. Chromatin immunoprecipitation (ChIP) of plant transcription factors followed by sequencing (ChIP-SEQ) or hybridization to whole genome arrays (ChIP-CHIP). *Nat. Protoc.* 5:457–72.

74. Quail, M. A., I. Kozarewa, F. Smith, A. Scally, P. J. Stephens, R. Durbin, H. Swerdlow, and D. J. Turner. 2008. A large genome center's improvements to the Illumina sequencing system. *Nat. Methods* 5:1005–10.

75. Zhou, H. W., D. F. Li, N. F. Tam, X. T. Jiang, H. Zhang, H. F. Sheng, J. Qin, X. Liu, and F. Zou. 2011. BIPES, a cost-effective high-throughput method for assessing microbial diversity. *Isme J.* 5:741–9.

76. O'Farrell, P. H. 1975. High resolution two-dimensional electrophoresis of proteins. *J. Biol. Chem.* 250:4007–21.

77. Guttman, A., Z. Csapo, and D. Robbins. 2002. Rapid two-dimensional analysis of proteins by ultra-thin layer gel electrophoresis. *Proteomics* 2:469–74.

78. Lamarcq, L., P. Lorimier, A. Negoescu, F. Labat-Moleur, I. Durrant, and E. Drumbillar 1995. Comparison of common bio- and chemiluminescent reagents for *in situ* detection of antigens and nucleic acids. *J. Biolumin. Chemilumin.* 10:247–56.

79. Samineni, S., S. Parvataneni, C. Kelly, V. Gangur, W. Karmaus, and K. Brooks. 2006. Optimization, comparison, and application of colorimetric vs. chemiluminescence based indirect sandwich ELISA for measurement of human IL-23. *J. Immunoassay Immunochem.* 27:183–93.

80. Campbell, G. T. and A. S. Bhatnagar. 1976. Simultaneous visualization by light microscopy of two pituitary hormones in a single tissue section using a combination of indirect immunohistochemical methods. *J. Histochem. Cytochem.* 24:448–52.

81. Soderberg, O., M. Gullberg, M. Jarvius, K. Ridderstrale, K. J. Leuchowius, J. Jarvius, K. Wester, P. Hydbring, F. Bahram, L. G. Larsson, and U. Landegren. 2006. Direct observation of individual endogenous protein complexes *in situ* by proximity ligation. *Nat. Methods* 3:995–1000.

82. Melin, J., J. Jarvius, C. Larsson, O. Soderberg, U. Landegren, and M. Nilsson. 2008. Ligation-based molecular tools for lab-on-a-chip devices. *Nat. Biotechnol.* 25:42–8.

83. Jarvius, M., J. Paulsson, I. Weibrecht, K. J. Leuchowius, A. C. Andersson, C. Wahlby, M. Gullberg, J. Botling, T. Sjoblom, B. Markova, A. Ostman, U. Landegren, and O. Soderberg. 2007. *In situ* detection of phosphorylated platelet-derived growth factor receptor beta using a generalized proximity ligation method. *Mol. Cell. Proteomics* 6:1500–9.

84. Smith, V., G. J. Wirth, H. H. Fiebig, and A. M. Burger. 2008. Tissue microarrays of human tumor xenografts: Characterization of proteins involved in migration and angiogenesis for applications in the development of targeted anticancer agents. *Cancer Genomics Proteomics* 5:26–73.

85. Marx, A., P. Simon, R. Simon, M. Mirlacher, J. R. Izbicki, E. Yekebas, J. T. Kaifi, L. Terracciano, and G. Sauter. 2008. AMACR expression in colorectal cancer is associated with left-sided tumor localization. *Virchows Arc.* 453:243–8.

86. Dahl, E., F. Wiesmann, M. Woenckhaus, R. Stoehr, P. J. Wild, J. Veeck, R. Knuchel, E. Klopocki, G. Sauter, R. Simon, W. F. Wieland, B. Walter, S. Denzinger, A. Hartmann, and C. G. Hammerschmied. 2007. Frequent loss of SFRP1 expression in multiple human solid tumours: Association with aberrant promoter methylation in renal cell carcinoma. *Oncogene* 26:5680–91.

87. Kononen, J., L. Bubendorf, A. Kallioniemi, M. Barlund, P. Schraml, S. Leighton, J. Torhorst, M. J. Mihatsch, G. Sauter, and O. P. Kallioniemi. 1998. Tissue microarrays for high-throughput molecular profiling of tumor specimens. *Nat. Med.* 4:844–7.

88. Engvall, E. and P. Perlman. 1971. Enzyme-linked immunosorbent assay (ELISA). Quantitative assay of immunoglobulin G. *Immunochemistry* 8:871–4.

89. Wang, Y., D. Wei, H. Yang, Y. Yang, W. Xing, Y. Li, and A. Deng. 2009. Development of a highly sensitive and specific monoclonal antibody-based enzyme-linked immunosorbent assay (ELISA) for detection of Sudan I in food samples. *Talanta* 77:1783–9.

90. Backen, A. C., J. Cummings, C. Mitchell, G. Jayson, T. H. Ward, and C. Dive. 2009. "Fit for purpose" validation of SearchLight multiplex ELISAs of angiogenesis for clinical trial use. *J. Immunol. Methods* 342:106–14.

91. Al-Dujaili, E. A., L. J. Mullins, M. A. Bailey, and C. J. Kenyon. 2009. Development of a highly sensitive ELISA for aldosterone in mouse urine: Validation in physiological and pathophysiological states of aldosterone excess and depletion. *Steroids* 74:456–62.

92. Gonzalez-Gomez, D., D. Cohen, J. A. Dickerson, X. Chen, F. Canada-Canada, and N. J. Dovichi. 2009. Improvement in protein separation in Barretts esophagus samples using two-dimensional capillary electrophoresis analysis in presence of cyclodextrins as buffer additives. *Talanta* 78:193–8.

93. Huang, T. L. and M. Richards. 1997. Development of a high-performance capillary isoelectric focusing technique with application to studies of microheterogeneity in chicken conalbumin. *J. Chromatogr. A.* 757:247–53.

94. Mario, N., B. Baudin, C. Aussel, and J. Giboudeau. 1997. Capillary isoelectric focusing and high-performance cation-exchange chromatography compared for qualitative and quantitative analysis of hemoglobin variants. *Clin. Chem.* 43: 2137–42.

95. Rainer, M., N. U. Muhammad, C. W. Huck, I. Feuerstein, R. Bakry, L. A. Huber, D. T. Gjerde, X. Zou, H. Qian, X. Du, F. Wei-Gang, Y. Ke, and G. K. Bonn. 2006. Ultra-fast mass fingerprinting by high-affinity capture of peptides and proteins on derivatized poly (glycidyl methacrylate/divinylbenzene) for the analysis of serum and cell lysates. *Rapid Commun. Mass Spectrom.* 20:2954–60.

96. Hillenkamp, F., M. Karas, R. C. Beavis, and B. T. Chait. 1991. Matrix-assisted laser desorption/ionization mass spectrometry of biopolymers. *Anal. Chem.* 63:1193A–1203A.

97. MacBeath, G. 2002. Protein microarrays and proteomics. *Nat. Genet.* 32(Suppl.):526–32.

98. Kusnezow, W. and J. D. Hoheisel. 2002. Antibody microarrays: Promises and problems. *Biotechniques* (Suppl.):14–23.

99. Zhou, H., J. D. Watts, and R. Aebersold. 2001. A systematic approach to the analysis of protein phosphorylation. *Nat. Biotechnol.* 19:375–8.

100. Ong, S. E., B. Blagoev, I. Kratchmarova, D. B. Kristensen, H. Steen, A. Pandey, and M. Mann. 2002. Stable isotope labeling by amino acids in cell culture, SILAC, as a simple and accurate approach to expression proteomics. *Mol. Cell Proteomics* 1:376–86.

101. Prokhorova, T. A., K. T. Rigbolt, P. T. Johansen, J. Henningsen, I. Kratchmarova, M. Kassem, and B. Blagoev. 2009. SILAC-labeling and quantitative comparison of the membrane proteomes of self-renewing and differentiating human embryonic stem cells. *Mol. Cell Proteomics* 8:959–70.

102. Abdo, M., B. Irving, P. Hudson, and H. Zola 2005. Development of a cluster of differentiation antibody-based protein microarray. *J. Immunol. Methods* 305:3–9.

103. Haab, B. B., M. J. Dunham, and P. O. Brown. 2001. Protein microarrays for highly parallel detection and quantitation of specific proteins and antibodies in complex solutions. *Genome Biol.* 2:r0004–r0004.13.

104. Miki, Y., T. Suzuki, C. Tazawa, Y. Yamaguchi, K. Kitada, S. Honma, T. Moriya, H. Hirakawa, D. B. Evans, S. Hayashi, N. Ohuchi, and H. Sasano. 2007. Aromatase localization in human breast cancer tissues: Possible interactions between intratumoral stromal and parenchymal cells. *Cancer Res.* 67:3945–54.

105. Brown, J. D., S. Dutta, K. Bharti, R. F. Bonner, P. J. Munson, I. B. Dawid, A. L. Akhtar, I. F. Onojafe, R. P. Alur, J. M. Gross, J. P. Hejtmancik, X. Jiao, W. Y. Chan, and B. P. Brooks. 2009. Expression profiling during ocular development identifies 2 Nlz genes with a critical role in optic fissure closure. *P. Natl. Acad. Sci. USA* 105:1461467.

106. Xu, B. J., J. Li, R. D. Beauchamp, Y. Shyr, M. Li, M. K. Washington, T. J. Yeatman, R. H. Whitehead, R. J. Coffey, and R. M. Caprioli. 2009. Identification of early intestinal neoplasia protein biomarkers using laser capture microdissection and MALDI MS. *Mol. Cell Proteomics* 8:936–45.

107. Jacob, K., C. Sollier, and N. Jabado. 2007. Circulating tumor cells: Detection, molecular profiling and future prospects. *Expert Rev. Proteomics* 4:741–56.

108. Bujold, E., R. Romero, J. P. Kusanovic, O. Erez, F. Gotsch, T. Chaiworapongsa, R. Gomez, J. Espinoza, E. Vaisbuch, Y. Mee Kim, et al. 2008. Proteomic profiling of amniotic fluid in preterm labor using two-dimensional liquid separation and mass spectrometry. *J. Matern. Fetal Neonatal. Med.* 21:697–713.

109. Galamb, O., S. Spisak, F. Sipos, K. Toth, N. Solymosi, B. Wichmann, T. Krenacs, G. Valcz, Z. Tulassay, and B. Molnar. 2010. Reversal of gene expression changes in the colorectal normal-adenoma pathway by NS398 selective COX2 inhibitor. *Brit. J. Cancer* 102:765–73.

110. Stott, S. L., C. H. Hsu, D. I. Tsukrov, M. Yu, D. T. Miyamoto, B. A. Waltman, S. M. Rothenberg, A. M. Shah, M. E. Smas, et al. 2010. Isolation of circulating tumor cells using a microvortex-generating herringbone-chip. *Proc. Natl. Acad. Sci. USA* 107:18397.

111. Stott, S. L., R. J. Lee, S. Nagrath, M. Yu, D. T. Miyamoto, L. Ulkus, E. J. Inserra, M. Ulman, S. Springer, Z. Nakamura, A. L. Moore, D. I. Tsukrov, M. E. Kempner, D. M. Dahl, C. L. Wu, A. J. Iafrate, M. R. Smith, R. G. Tompkins, L. V. Sequist, M. Toner, D. A. Haber, and S. Maheswaran. 2010. Isolation and characterization of circulating tumor cells from patients with localized and metastatic prostate cancer. *Sci. Transl. Med.* 2:25ra23.

112. Lizardi, P. M., V. Mahdavi, D. Shields, and G. Candelas. 1979. Discontinuous translation of silk fibroin in a reticulocyte cell-free system and in intact silk gland cells. *P. Natl. Acad. Sci. USA* 76:6211–5.

113. Kusnezow, W. and J. D. Hoheisel. 2003. Solid supports for microarray immunoassays. *J. Mol. Recognit.* 16:165–76.
114. Nijdam, A. J., M. Ming-Cheng Cheng, D. H. Geho, R. Fedele, P. Herrmann,
K. Killian, V. Espina, E. F. Petricoin, 3rd, L. A. Liotta, and M. Ferrari. 2007.
Physicochemically modified silicon as a substrate for protein microarrays.
Biomaterials 28:550–8.
115. Pavlickova, P., N. M. Jensen, H. Paul, M. Schaeferling, C. Giammasi, M.
Kruschina, W. D. Du, M. Theisen, M. Ibba, F. Ortigao, and D. Kambhampati.
2002. Antibody detection in human serum using a versatile protein chip
platform constructed by applying nanoscale self-assembled architectures on
gold. *J. Proteome Res.* 1:227–31.
116. Derwinska, K., L. A. Gheber, and C. Preininger. 2007. A comparative analysis
of polyurethane hydrogel for immobilization of IgG on chips. *Anal. Chim.
Acta* 592:138.
117. Fixe, F., M. Dufva, P. Telleman, and C. B. Christensen. 2004. Functionalization
of poly(methyl methacrylate) (PMMA) as a substrate for DNA microarrays.
Nucleic Acids Res. 32:e9.
118. Carion, O., V. Souplet, C. Olivier, C. Maillet, N. Medard, O. El-Mahdi, J. O.
Durand, and O. Melnyk. 2007. Chemical micropatterning of polycarbonate for site-specific peptide immobilization and biomolecular interactions.
ChemBioChem. 8:315–22.
119. Li, Y., Z. Wang, L. M. Ou, and H. Z. Yu. 2007. DNA detection on plastic:
Surface activation protocol to convert polycarbonate substrates to biochip
platforms. *Anal. Chem.* 79:426–33.
120. Afanassiev, V., V. Hanemann, and S. Wolfl. 2000. Preparation of DNA and
protein micro arrays on glass slides coated with an agarose film. *Nucleic
Acids Res.* 28:E66.
121. Liu, Y., C. M. Li, W. Hu, and Z. Lu. 2009. High performance protein microarrays based on glycidyl methacrylate-modified polyethylene terephthalate
plastic substrate. *Talanta* 77:1165–71.
122. Zhu, H. and M. Snyder. 2001. Protein arrays and microarrays. *Curr. Opin.
Chem. Biol.* 5:40–5.
123. Joos, T. O., M. Schrenk, P. Hopfl, K. Kroger, U. Chowdhury, D. Stoll, D.
Schorner, M. Durr, K. Herick, S. Rupp, K. Sohn, and H. Hammerle. 2000. A
microarray enzyme-linked immunosorbent assay for autoimmune diagnostics. *Electrophoresis* 21:2641–50.
124. Quintana, F. J., M. F. Farez, V. Viglietta, A. H. Iglesias, Y. Merbl, G. Izquierdo,
M. Lucas, A. S. Basso, S. J. Khoury, C. F. Lucchinetti, I. R. Cohen, and H. L.
Weiner. 2008. Antigen microarrays identify unique serum autoantibody signatures in clinical and pathologic subtypes of multiple sclerosis. *Proc. Natl.
Acad. Sci. USA* 105:18889–94.
125. Patwa, T. H., Y. Wang, D. M. Simeone, and D. M. Lubman. 2008. Enhanced
detection of autoantibodies on protein microarrays using a modified protein
digestion technique. *J. Proteome Res.* 7:2553–61.
126. Gulmann, C., K. M. Sheehan, R. M. Conroy, J. D. Wulfkuhle, V. Espina, M. J.
Mullarkey, E. W. Kay, L. A. Liotta, and E. F. Petricoin, III. 2009. Quantitative
cell signalling analysis reveals down-regulation of MAPK pathway activation in colorectal cancer. *J. Pathol.* 218:514–9.

127. Chatterjee, M., J. Wojciechowski, and M. A. Tainsky. 2009. Discovery of antibody biomarkers using protein microarrays of tumor antigens cloned in high throughput. *Methods Mol. Biol.* 520:21–38.
128. Cretich, M., G. Di Carlo, C. Giudici, S. Pokoj, I. Lauer, S. Scheurer, and M. Chiari. 2009. Detection of allergen specific immunoglobulins by microarrays coupled to microfluidics. *Proteomics* 9:2098–107.
129. Bartel, D. 2004. MicroRNAs: Genomics, biogenesis, mechanism, and function. *Cell* 107:823–6.
130. Dugas, D. V. and B. Bartel. 2004. MicroRNA regulation of gene expression in plants. *Curr. Opin. Plant Biol.* 7:5120.
131. Abbott, A. L., E. Alvarez-Saavedra, E. A. Miska, N. C. Lau, D. P. Bartel, H. R. Horvitz, and V. Ambros. 2005. The let-7 microRNA family members mir-48, mir-84, and mir-241 function together to regulate developmental timing in *Caenorhabditis elegans*. *Dev. Cell* 9:403–14.
132. Ambros, V. 2001. MicroRNAs: Tiny regulators with great potential. *Cell* 107:823–6.
133. Ambros, V. 2003. MicroRNA pathways in flies and worms: Growth, death, fat, stress, and timing. *Cell* 113:673–6.
134. Ambros, V. and X. Chen. 2007. The regulation of genes and genomes by small RNAs. *Development* 134:1635–41.
135. Lee, R., R. Feinbaum, and V. Ambros. 2004. A short history of a short RNA. *Cell* 116(Suppl.):S89–92.
136. Lee, R. C. and V. Ambros. 2001. An extensive class of small RNAs in *Caenorhabditis elegans*. *Science* 294:864.
137. Khanna, A. and S. Stamm. 2010. Regulation of alternative splicing by short non-coding nuclear RNAs. *RNA Biol.* 7:480–5.
138. Wang, H., R. A. Ach, and B. Curry. 2007. Direct and sensitive miRNA profiling from low-input total RNA. *RNA* 13:151–9.
139. Biscontin, A., S. Casara, S. Cagnin, L. Tombolan, A. Rosolen, G. Lanfranchi, and C. De Pitta. 2010. New miRNA labeling method for bead-based quantification. *BMC Mol. Biol.* 11:44.
140. Gaarz, A., S. Debey-Pascher, S. Classen, D. Eggle, B. Gathof, J. Chen, J. B. Fan, T. Voss, J. L. Schultze, and A. Staratschek-Jox. 2010. Bead array-based microRNA expression profiling of peripheral blood and the impact of different RNA isolation approaches. *J. Mol. Diagn.* 12:335–44.
141. Nana-Sinkam, S. P. and C. M. Croce. 2010. MicroRNA dysregulation in cancer: Opportunities for the development of microRNA-based drugs. *IDrugs* 13:843–6.
142. Chen, C., D. A. Ridzon, A. J. Broomer, Z. Zhou, D. H. Lee, J. T. Nguyen, M. Barbisin, N. L. Xu, V. R. Mahuvakar, M. R. Andersen, K. Q. Lao, K. J. Livak, and K. J. Guegler. 2005. Real-time quantification of microRNAs by stem-loop RT-PCR. *Nucleic Acids Res.* 33:179.
143. Feng, J., K. Wang, X. Liu, S. Chen, and J. Chen. 2009. The quantification of tomato microRNAs response to viral infection by stem-loop real-time RT-PCR. *Gene.* 437:14–21.
144. Cao, P. and J. T. Stults. 2000. Mapping the phosphorylation sites of proteins using on-line immobilized metal affinity chromatography/capillary electrophoresis/electrospray ionization multiple stage tandem mass spectrometry. *Rapid. Commun. Mass Spectrom.* 14:1600–6.

145. Figeys, D. 2003. Proteomics in 2002: A year of technical development and wide-ranging applications. *Anal. Chem.* 75:2891–905.
146. Melle, C., G. Ernst, B. Schimmel, A. Bleul, S. Koscielny, A. Wiesner, R. Bogumil, U. Moller, D. Osterloh, K. J. Halbhuber, and F. von Eggeling. 2004. A technical triade for proteomic identification and characterization of cancer biomarkers. *Cancer Res.* 64:4099–104.
147. Ueda, M., Y. Misumi, M. Mizuguchi, M. Nakamura, T. Yamashita, Y. Sekijima, K. Ota, S. Shinriki, H. Jono, S. I. Ikeda, O. B. Suhr, and Y. Ando. 2009. SELDI-TOF Mass spectrometry evaluation of variant transthyretins for diagnosis and pathogenesis of familial amyloidotic polyneuropathy. *Clin. Chem.* 55:1223–7.
148. Proll, G., L. Steinle, F. Proll, M. Kumpf, B. Moehrle, M. Mehlmann, and G. Gauglitz. 2007. Potential of label-free detection in high-content-screening applications. *J. Chromatogr. A.* 1161:2–8.
149. Buijs, J. and G. C. Franklin. 2005. SPR-MS in functional proteomics. *Brief Funct. Genomic Proteomics.* 4:39–47.
150. Borch, J. and P. Roepstorff. 2004. Screening for enzyme inhibitors by surface plasmon resonance combined with mass spectrometry. *Anal. Chem.* 76:5243–8.
151. Zhukov, A., M. Schurenberg, O. Jansson, D. Areskoug, and J. Buijs. 2004. Integration of surface plasmon resonance with mass spectrometry: Automated ligand fishing and sample preparation for MALDI MS using a Biacore 3000 biosensor. *J. Biomol. Technol.* 15:112–9.
152. McLafferty, F. W., E. K. Fridriksson, D. M. Horn, M. A. Lewis, and R. A. Zubarev. 1999. Techview: Biochemistry. Biomolecule mass spectrometry. *Science* 284:1289–90.
153. Vaisocherova, H., Z. Zhang, W. Yang, Z. Cao, G. Cheng, A. D. Taylor, M. Piliarik, J. Homola, and S. Jiang. 2008. Functionalizable surface platform with reduced nonspecific protein adsorption from full blood plasma-material selection and protein immobilization optimization. *Biosens. Bioelectron.* 24:1924–30.
154. Huang, Y. Y., H. Y. Hsu, and C. J. Huang. 2007. A protein detection technique by using surface plasmon resonance (SPR) with rolling circle amplification (RCA) and nanogold-modified tags. *Biosens. Bioelectron.* 22:980–5.
155. Yurkovetsky, Z. R., J. M. Kirkwood, H. D. Edington, A. M. Marrangoni, L. Velikokhatnaya, M. T. Winans, E. Gorelik, and A. E. Lokshin. 2007. Multiplex analysis of serum cytokines in melanoma patients treated with interferon-alpha2b. *Clin. Cancer Res.* 13:2422–8.
156. Ellington, A. D. and J. W. Szostak. 1990. *In vitro* selection of RNA molecules that bind specific ligands. *Nature* 346:818–22.
157. Tuerk, C. and L. Gold. 1990. Systematic evolution of ligands by exponential enrichment: RNA ligands to bacteriophage T4 DNA polymerase. *Science* 249: 505–10.
158. Robertson, D. L. and G. F. Joyce. 1990. Selection *in vitro* of an RNA enzyme that specifically cleaves single-stranded DNA. *Nature* 344:467–8.
159. Yang, Y., D. Yang, H. J. Schluesener, and Z. Zhang. 2007. Advances in SELEX and application of aptamers in the central nervous system. *Biomol. Eng.* 24:583–92.

160. Stoltenburg, R., C. Reinemann, and B. Strehlitz. 2007. SELEX—A (r)evolutionary method to generate high-affinity nucleic acid ligands. *Biomol. Eng.* 24:381–403.

161. Chhabra, R., J. Sharma, Y. Ke, Y. Liu, S. Rinker, S. Lindsay, and H. Yan. 2007. Spatially addressable multiprotein nanoarrays templated by aptamer-tagged DNA nanoarchitectures. *J. Am. Chem. Soc.* 129:10304–5.

162. Blank, M. and M. Blind. 2005. Aptamers as tools for target validation. *Curr. Opin. Chem. Biol.* 9:336–42.

163. Cheng, A. K., D. Sen, and H. Z. Yu. 2009. Design and testing of aptamer-based electrochemical biosensors for proteins and small molecules. *Bioelectrochemistry* 77:1–12.

164. Mairal, T., V. C. Ozalp, P. Lozano Sanchez, M. Mir, I. Katakis, and C. K. O'Sullivan. 2008. Aptamers: Molecular tools for analytical applications. *Anal. Bioanal. Chem.* 390:989–1007.

165. Ratner, D. M. and P. H. Seeberger. 2007. Carbohydrate microarrays as tools in HIV glycobiology. *Curr. Pharm. Design* 13:173–83.

166. Patwa, T., C. Li, D. M. Simeone, and D. M. Lubman. 2010. Glycoprotein analysis using protein microarrays and mass spectrometry. *Mass Spectrom. Rev.* 29:830–44.

167. Ianni, M., M. Manerba, G. Di Stefano, E. Porcellini, M. Chiappelli, I. Carbone, and F. Licastro. 2010. Altered glycosylation profile of purified plasma ACT from Alzheimer's disease. *Immun. Ageing.* 7(Suppl.):S6.

168. Rosenfeld, R., H. Bangio, G. J. Gerwig, R. Rosenberg, R. Aloni, Y. Cohen, Y. Amor, I. Plaschkes, J. P. Kamerling, and R. B. Maya. 2007. A lectin array-based methodology for the analysis of protein glycosylation. *J. Biochem. Biophys. Methods* 70:415–26.

169. Coullerez, G., P. H. Seeberger, and M. Textor. 2006. Merging organic and polymer chemistries to create glycomaterials for glycomics applications. *Macromol. Biosci.* 6:634–47.

170. Zhao, J., T. H. Patwa, W. Qiu, K. Shedden, R. Hinderer, D. E. Misek, M. A. Anderson, D. M. Simeone, and D. M. Lubman. 2007. Glycoprotein microarrays with multi-lectin detection: Unique lectin binding patterns as a tool for classifying normal, chronic pancreatitis and pancreatic cancer sera. *J. Proteome Res.* 6:1864–74.

171. Gunasekera, U. A., Q. A. Pankhurst, and M. Douek. 2009. Imaging applications of nanotechnology in cancer. *Target Oncol.* 4:169–81.

172. Nie, S., Y. Xing, G. J. Kim, and J. W. Simons. 2007. Nanotechnology applications in cancer. *Annu. Rev. Biomed. Eng.* 9:257–88.

173. Seigneuric, R., L. Markey, D. S. Nuyten, C. Dubernet, C. T. Evelo, E. Finot, and C. Garrido. 2010. From nanotechnology to nanomedicine: Applications to cancer research. *Curr. Mol. Med.* 10:640–52.

174. Johnson, L., A. Gunasekera, and M. Douek. 2010. Applications of nanotechnology in cancer. *Discov. Med.* 9:374–9.

175. Pilobello, K. T., D. E. Slawek, and L. K. Mahal. 2007. A ratiometric lectin microarray approach to analysis of the dynamic mammalian glycome. *P. Natl. Acad. Sci. USA* 104:11534–9.

176. Shin, I., S. Park, and M. R. Lee. 2005. Carbohydrate microarrays: An advanced technology for functional studies of glycans. *Chemistry* 11:2894–901.

177. Wang, H., Y. Zhang, X. Yuan, Y. Chen, and M. Yan. 2011. A universal protocol for photochemical covalent immobilization of intact carbohydrates for the preparation of carbohydrate microarrays. *Bioconjug. Chem.* 22:26–32.
178. Nan, G., H. Yan, G. Yang, Q. Jian, C. Chen, and Z. Li. 2009. The hydroxyl-modified surfaces on glass support for fabrication of carbohydrate microarrays. *Curr. Pharm. Biotechnol.* 10:138–46.
179. Thompson, S. M., D. G. Fernig, E. C. Jesudason, P. D. Losty, E. M. van de Westerlo, T. H. van Kuppevelt, and J. E. Turnbull. 2009. Heparan sulfate phage display antibodies identify distinct epitopes with complex binding characteristics: Insights into protein binding specificities. *J. Biol. Chem.* 284:35621–31.
180. Thompson, S. M., M. G. Connell, D. G. Fernig, G. B. Ten Dam, T. H. van Kuppevelt, J. E. Turnbull, E. C. Jesudason, and P. D. Losty. 2007. Novel phage display antibodies identify distinct heparan sulfate domains in developing mammalian lung. *Pediatr. Surg. Int.* 23:411–7.

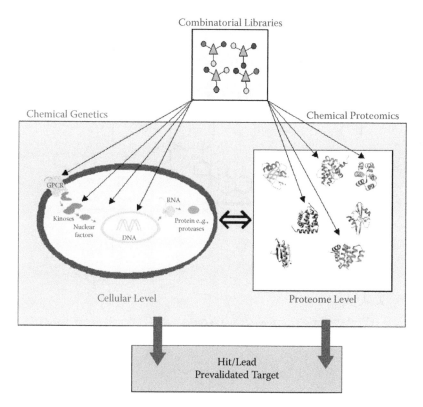

Figure 1.1 Overview of the utilization of small molecules in chemical genetics/proteomics.

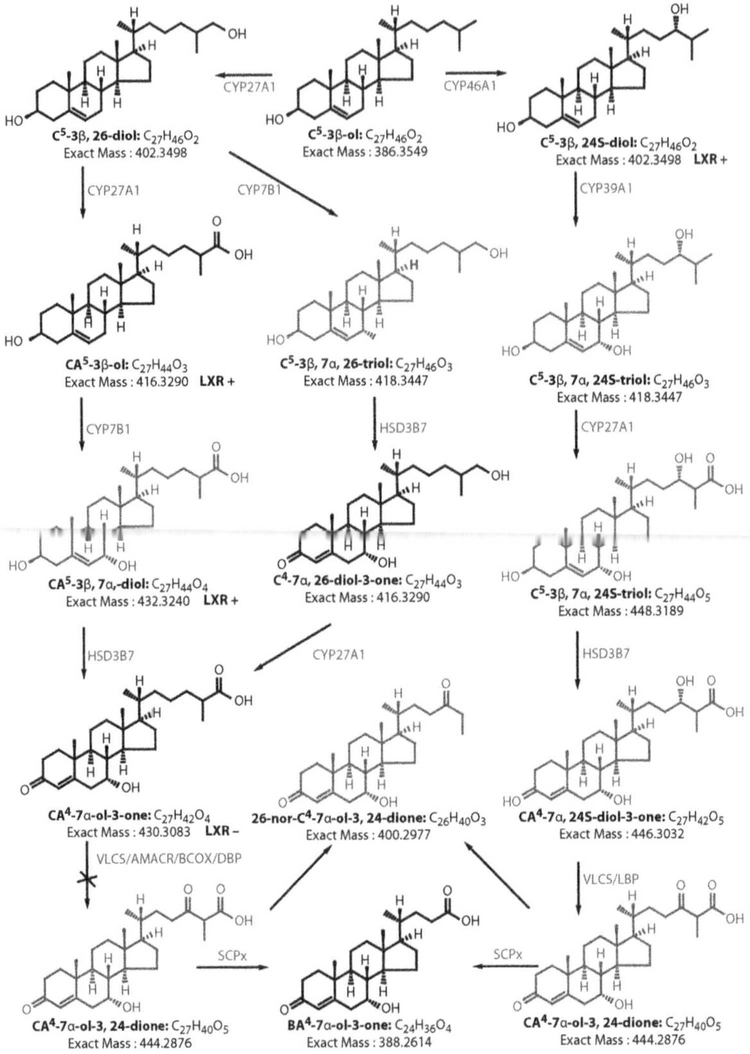

Figure 2.17 Suggested pathway of cholesterol metabolism in human brain based on the identification of metabolites in CSF. The left hand branch constitutes the initial steps of the acidic pathway of bile acid biosynthesis. 26-Hydroxycholesterol may be synthesized in brain via a reaction catalyzed by CYP27A1 or imported into brain from the circulation (Heverin, Meaney, Lütjohann, et al. [60]). Thus, the ultimate product of this pathway 7α-hydroxy-3-oxocholest-4-en-26-oic may be derived from brain synthesized or imported sterols. The right hand branch of the pathway represents the initial steps of the 24-hydroxylase pathway of bile acid biosynthesis. The initial enzyme CYP46A1 is uniquely expressed in nervous tissue (Lund, Guileyardo, and Russell [61]).

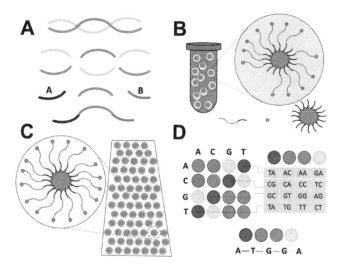

Figure 3.3 SOLiD technology. Fragmentation of a template DNA and specific adapter (A and B) ligation (A). Emulsion PCR in individual microreactors, which resulted in clonal amplification of each template molecule (B). The surface of the sequencing slide after loading the amplified template (C). The basis of color-coding (D).

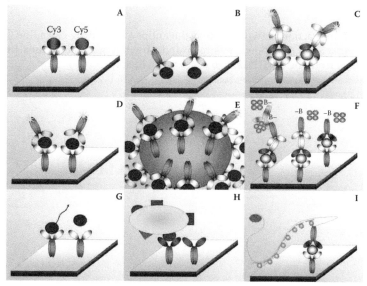

Figure 3.6 Summary diagram of protein microchip labeling/detection techniques and main areas, including: (A) Direct labeling. (B) Reversed-phase array (RPA). (C) Labeling with tertiary antibody. (D) Sandwich labeling. (E) Bead-based array. (F) Biotin-labeled antibody. (G) Label-free method for mass spectrometry. (H) Whole-cell array. (I) Rolling Circle Amplification (RCA).

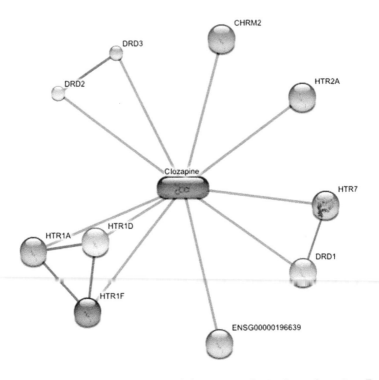

Figure 6.3 Interaction network around the antipsychotic drug clozapine. Drug–target interactions are shown in green and target–target interactions in blue. (Data taken from the STITCH2 Web site: http://stitch.embl.de/.)

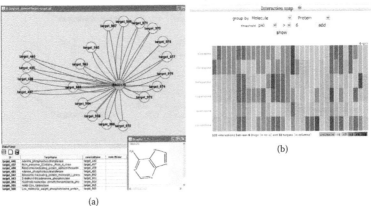

Figure 6.4 (a) Displaying the drug-target interaction network around adenine (DB00173) with DrugViz. (b) Interaction map of clozapine and five related compounds with 32 targets, generated with iPHACE. (From Garcia-Serna, R. et al., 2010, iPHACE: Integrative Navigation in Pharmacological Space, *Bioinformatics* 26:985–986.)

chapter four

Genomic and proteomic biomarkers in the drug R&D process

László Takács, Anna Debreceni, and István Kurucz

Contents

4.1 Discovery and development of biomarkers

Biomolecular biomarkers are used often in the diagnosis of diseases, at some stages in the small molecule drug R&D process and during the therapeutic use of marketed medicines [1]. Throughout the process, the use of biomarkers is intended to secure the highest level of efficiency and

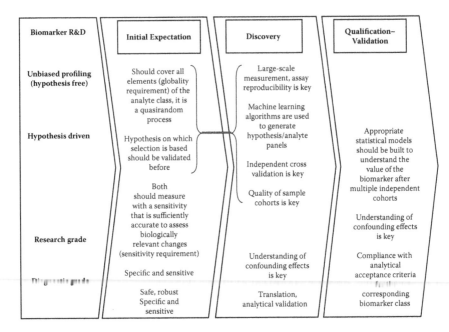

Figure 4.1 General scheme of the biomolecular–biomarker R&D process.

safety possible for a given drug, that is, during the drug discovery phase to select the best chemical entity that hits the target or during the drug development phase, to define the target patient group which will respond to treatment.

Biomarker discovery is an applied form of genome research. In this definition, the term *genome research* is extended and it includes proteomics, metabolomics, and lipidomics, as disease associated or normal variation of the respective analytes is the consequence of genome function. In this context, environmental impact is not considered as it is outside of address-able biology. In short, the ultimate, unbiased biomarker discovery process is random exploration, a search for measurable variables (see Figure 4.1) to find reproducibly changing patterns that signal new paradigms [2] and lead to new understanding or can be linked to hypotheses set before the experiment. By the nature of the process, all naturally occurring biomol-ecules can be biomarkers, and these represent the subject of this chapter. In addition to naturally occurring biomolecules, imaging technologies, such as ultrasound, X-ray, computed tomography (CT), positron emission tomography (PET), and nuclear magnetic resonance (NMR) imaging as well as physiological, morphometric (i.e., body mass index), histometric, psychological, and cognitive testing are also rich sources of biomarkers, however none of these markers are addressed in this review.

In humans, DNA, mRNA, protein, and metabolites are the basic classes of biomolecules. It is useful to address these separately, as technologies used to measure and profile the different classes represent different challenges but in the context of this review, the group is treated as one, with few exceptions.

Hypothesis-free biomarker discovery should fulfill the *"globality"* requirement, thus all elements of the targeted analyte type (i.e., mRNAs transcribed from all genes) should be measurable with the required sensitivity in any single experiment. The *globality in vitro* requirement is fulfilled by DNA sequencing technologies, as with the advent of next-generation sequencing technologies, even the most complex individual genomes, such as human, can be sequenced and analyzed in the specialized laboratory setting [3]. Transcript (mRNA) profiling via microchips and *in vitro message counting* by deep cDNA sequencing also satisfies the *globality* criterion; however, today there is no proteome profiling and metabolome profiling technology that fulfills the *globality* requirement [4]. Nevertheless, shotgun mass spectrometry is frequently used today despite its inability to cover all targeted analytes (i.e., molecules that are not reproducibly ionizable) in one experiment [5]. Other key requirements are *reproducibility* and *sensitivity*: technologies applied in biomarker discovery should be reproducible (i.e., detection of a single nucleotide polymorphism in populations of hundreds of thousands of individuals) and sensitive enough to detect quantitative and qualitative changes, which occur during a biological response [6].

In hypothesis-driven discovery programs, often a list of candidate markers are compiled using the current knowledge base. The list of candidates may be compiled with advanced bioinformatics data mining tools (i.e., electronic interpretation of scientific papers), which could generate candidates that represent members of a probable pathway, either as expressed genes, or proteins, or the affected metabolites.

In either case, hypothesis driven or not, biomarker discovery generates large data sets. A challenging task then is to find the minimal panel of markers that provides the maximum information content with respect to an emerging new paradigm (e.g., a subgroup of drug-responding patients identified by mRNA profiling of cancer tissue) (see Oncotype DX™ from Genomic Health, etc.). As *a priori* knowledge does not exist, special algorithms are needed to find patterns inherently present in the data. Among others, these include machine learning algorithms, such as decision tree learning, neuronal networks, and support vector machines [7]. The results are often sensitive to special *hidden* features of the samples, for example, sample collection procedures; thus independent and multiple cycles of validation are of key importance already at the initial discovery phase of biomarkers. Without independent validation, biomarkers cannot be declared as *qualified* or *validated*. Validation is a continuous work [8], a series of experiments, in which all possible confounding factors (e.g., blood sampling procedures) that could produce similar results to the one observed and presumed to be the consequence of the intervention (i.e., drug treatment) are eliminated. Research-grade biomarkers (for example, those used in animal experiments to declare the validity of a new drug target as relevant to human, but selected in an animal model) can be declared valid, after multiple cycles of testing many experimental conditions that mimic human confounding factors (age, sex, metabolic status, environmental conditions, etc.).

Research-grade biomarkers do not need to be robust and do not need to comply with analytical criteria set forth for *in vitro* diagnostics on the regulated market as these are generally run in an individual research laboratory setting, where experimental conditions are controlled, and sample collection and other logistical procedures are not variable. However, diagnostic-grade biomarkers will have to be translated to simple and robust assays, which will generate results with less variability due to operator-dependent or laboratory-dependent variations. Final qualification and validation can only be declared, especially in the case of diagnostic-grade biomarkers, after completing the analytical development and for multivariate panels, after building statistical models that determine the statistical power of any given application. An important element of the statistical model building is the disease prevalence, which may be very different in the context of a defined clinical trial or on the market after the release of the diagnostics.

Analytical validation for diagnostic-grade biomarkers is necessary because on one hand, on the regulated *in vitro* diagnostic (IVD) market, analytical performance has to comply with preset criteria (FDA, notified body, or regulatory organization), on the other hand, for the research use only (RUO) tests market, market penetration will be limited in the absence of credible analytical data. The approach to analytical validation is driven by assay category. From the analytical standpoint, assays can be quantitative and qualitative, or in this context, classification type. Quantitative assays are sometimes not possible to develop to the highest standards, that is, to Definitive Quantitative Assays (DQAs), which require defined reference standards and calibrators. If defined, analytically pure, characterized (with accurate quality control procedures), and reproducible standards and calibrators are not available, only Relative Quantitative Assays (RQAs) can be developed. With the use of biological standards, especially if the availability is not limited and storage does not present impenetrable issues, a high level of analytical requirements can be met even for the RQA. These are assay accuracy, precision, specificity and dilution linearity (between the lower limit of detection and the upper limit of detection), standard stability, and matrix stability [1,9].

The classification type of assays generate discrete binary responses (yes or no, i.e., the presence of a specific mutation); these assays need to meet the sensitivity and specificity criteria, but do not require performance range-driven standardization. It is important to note that in the routine practice, standards are provided with these tests for both of the binary outcomes.

Analytical validation requirements are variable for the different quantitative assay types, and the variability is often driven by the biological variation that needs to be covered by the assay. In general, the lower and the upper limits of detection have to be determined, and these values are important descriptors of the assay performance. Biological samples between the two values have to fit to the linear plot of the standards covering the range. Reproducibility is described by the variation coefficient (CV) and the assay recovery (RE) in % scale. Good assays have less than 15% CV and less than 20% RE. However, acceptance criteria described in documents released periodically by the FDA and notified bodies may be looser, that is, CV of 20% to 30%, and 67% of the measurements should have at least 20% RE [1,9]. A general scheme of the biomarker R&D process is shown in Figure 4.1.

4.2   Drug R&D and biomarkers

To highlight the use of biomarkers in the process of drug R&D, three aspects are the most helpful to keep in mind for each stage of the process.

1. What questions can be addressed by biomarkers?
2. What are the most appropriate types of biomarkers to address a particular question?
3. What is the most appropriate biomarker technology?

4.2.1   Disease identification and definition

The first step of the drug R&D process is planning the identification of the disease or the condition that will be addressed by the R&D project. Diagnostic-grade biomarkers are critical at this step. The biomarkers applied at this stage are validated to the highest possible level and are often FDA approved IVD tests. These markers are used to select the patients to whom the new drug will be given first in clinical trials and later therapeutically. As Westernized medicine evolves, the number of disease conditions are gradually becoming more and more defined. Thus, disease categories, such as diabetes or breast cancer, fall into entities that are becoming better defined in biomolecular terms, as in the example of type 1 (lack of insulin) and type 2 diabetes (insulin resistance) or HER2 positive and HER2 negative breast cancer. It is not surprising that effective therapies are going to be different for each particular condition.

Diagnostic-grade biomarkers then address the specific molecular aspects of disease conditions to ensure that the population targeted with a new drug is relatively homogeneous, which is a key factor to ensure drug efficacy. The types of markers studied at this planning and evaluation stage are early detection, stratification, and disease risk markers. The types of tests applied are robust and involve well-established technologies, that is, enzyme-linked immunosorbent assays (ELISA) using blood samples, or polymerase chain reaction (PCR) to define genetic variations that are present in a particular cancer type (Figure 4.1).

4.2.2   Target identification

Once the disease condition is selected and the projected market is attractive (based on the biomarker/IVD and imaging-driven diagnosis), the molecular drug target has to be identified. Generally, at the initial phase of this critical step, global molecular profiling technologies are used to find a *druggable* target that lies within an assumed disease critical pathway. Technologies and biomarkers used for the purpose are research grade, and often involve less validated and novel technologies, that is, new variants of the more widely applied *omics* technologies, like deep sequencing of cDNA to analyze gene expression profiles from specifically affected tissues, or shotgun mass spectrometry to profile the dynamics of the particular metabolome or proteome segment in cancer tissue. Following sophisticated statistical and informatics work, new *druggable* candidate genes are identified, which will have to be validated in painstaking *in silico* and *in vivo* experimentation before the process of drug discovery can progress. Naturally, in the case of drug R&D projects where the pathway and the target have already been identified, but are being addressed with new chemical entities, there is no need for discovery efforts to identify the target. In this situation, the focus is on finding special reporter assays indicating that the drug candidate actually hits the target.

4.2.3   Lead identification and definition

In the next step, for the lead compound identification and development, the requirement and often the outcome of the biomarker research is a specific assay that reports binding and affecting the selected target by a small chemical drug candidate. Robust reporter assays are often developed in order to test the interaction of candidate chemical entities with the drug target *in vitro* [10,11]. In the standard paradigm, high-throughput forms of the robust assays are applied for screening small molecular libraries for the identification of lead candidates. Biomarker technology requirements are highly specialized and include considerations related to efficient robotic applications of testing recombinant forms of the target protein, or sophisticated cell-based assays of genetically modified cell lines. Paradigms that deal with *in silico* model-based screens are also popular. Here, smaller scale experimentation addresses the validation of the *in silico* hits.

4.2.4 Animal models—Translational research

Once lead molecules are identified, in the subsequent step, these will have to be tested in animal models of the disease with the toolkit for translational research. A question often asked is: does the drug candidate hit its supposed biological target in such a way that we expect it to and with the best results and the least toxicity? Because this question is often approached with the use of complex *in vitro* or animal models, the biomarkers are called *in vitro* or *target biomarkers*, respectively. Gene manipulation of experimental animals is widespread, so we can test the supposed molecular mechanisms of human diseases using transgenic mice harboring the human gene target [12]. Markers with prognostic and predictive value in human application often are results of the biomarker discovery effort at this stage.

At this stage, a particularly important factor is that it has to be formally proven that the drug candidate actually hits the animal equivalent of the human drug target. As for particular molecular functions, the human genome often carries a redundant gene set in the form of gene isoforms; individual members of the isoform group are either present or absent in the animals and may have very different tissue distribution. The classification may be blurred because of the difficulties involved in separating sequence-based evolutionary relatedness from molecular redundancy within the group of related genes. Identification of *equivalency biomarkers* is key at this stage and it represents one of the most difficult tasks.

Animal models are also used to address pharmacodynamics and toxicology initially. A challenging task is finding the most appropriate biomarker detection method to follow the drug candidate and its decay in the complex organism. The technology is often mass spectrometry-based and involves lengthy analytical process development, which addresses the characterization of interactions of small molecules and their degradation products with blood proteins. Toxicology biomarkers are often standard enzymatic assays, reporting liver and kidney cell damage. These biomarkers and their laboratory assays are well established and widely used. While for target validation, the mouse is the most frequently used model, initial toxicity studies are often done first in rats, then in dogs. For correlation of toxicity with biological, disease, and target-related effects, there is a significant emerging need to find the equivalent of the target-specific biomarkers and often the efficacy-related markers as well in rats and dogs. With the advent of the complete genome sequencing of both species, this task is much easier now than it was before.

4.2.5   Phases 0–I

If candidate drugs are found to be efficacious and safe in animal models of the disease, human clinical testing starts. In the initial phases (Phases 0–I), during the dose testing, generally there is no biomarker other than the pharmacokinetic/pharmacodynamic (PK/PD), and toxicity markers within dedicated PK/PD programs. However, it has to be noted that initial efficacy of atorvastatin's (Lipitor®) was demonstrated in a Phase I trial setup [13] with an ultimate target and surrogate biomarker (LDL cholesterol); the trend is to copy this innovative approach. The motivation is to reduce the time for the development process. However, efficacy of new drugs generally are much lower than that of one of the most successful drug in history, Lipitor, and the demonstration of the required 20% to 30% improvement necessitates large and carefully defined populations for Phase II and Phase III testing. Also, the PK/PD team often faces issues related to the inability to achieve the assumed therapeutic effect due to low tolerance of the required dose or the unavailability of the drug for the target. Solving these issues is aided by the pharmacodynamic markers and often requires innovative approaches to detect various metabolized and active/inactive forms of the drug bound to albumin or other proteins in the blood and at the site of action.

4.2.6   Phases II–III

During the subsequent phases (Phases II–III), if the required effective dose can be achieved with good tolerance, the efficacy of the new drug is measured by the use of a few selected doses and defined clinical endpoints (CEs). In these phases of drug development, albeit often research grade, many biomarkers are used. Although research grade, the importance may be high, as Phase IIa, is the most critical step, because any change after this point will impact the economy of the development project in a critical way. Drugs entering Phase III and failing, even if the failure is due to a known and correctable cause are likely to be killed forever.

Three biomarker types are the most critical, these are: surrogate (disease management), stratification (companion), and early detection. The semantics is somewhat confusing as the same types of markers used during a drug trial may have different names compared to the use in clinical practice; here the naming used in clinical practice is in parentheses.

4.2.7 Surrogate biomarkers

The challenges stem from the fact that the desired outcome is often longer term than the practically achievable length of the clinical trial. Continuing the Lipitor example from above, correlation of LDL cholesterol level with cardiovascular morbidity risk, that is, getting heart attack, stroke, and mortality, was well established before the Lipitor trial. Thus, LDL cholesterol was accepted by the medical community as a surrogate marker of cardiovascular risk. At the same time Lipitor inhibits the key enzyme, HMG-CoA reductase, which is the rate-controlling enzyme (EC 1.1.1.88) of the mevalonate pathway, the metabolic pathway that produces cholesterol and other isoprenoids. Thus, in this unique case, there was a direct connection between three critical components, the drug (Lipitor), the target (HMG-CoA), and the surrogate marker (LDL cholesterol). The situation is often much more complex: the drug target can be hypothetical (efficacy in human has to be proven in the trial), pharmacological modulation of the target may not be easily measurable, and there could be no validated surrogate marker that will report drug efficacy with respect to the selected clinical outcome. Important and not too widely appreciated is that biomarkers can offer solutions, but as hypothesis-driven candidates may not work (in the lack of an apparent paradigm), unbiased hypothesis-free genomic, proteomic, and metabolomic profiling is often done at this stage. The genetic and gene expression data may even be voluntarily submitted to the FDA [14]. The data may include the types of genetic loci or gene expression profiles being explored by the pharmaceutical industry for pharmacogenomic testing, the test systems and techniques being employed, or the problems encountered in applying pharmacogenomic tests to drug development. In addition, these may include the scientific rationale for standardizing, naming, and characterizing the genes used on different genomic analysis platforms and for developing bioinformatics software programs used to evaluate pharmacogenomic data. For the public interest and largely for the discovery of surrogate markers in the future, to facilitate voluntary exploratory data submissions (VXDSs), the FDA has established a cross-center, the Interdisciplinary Pharmacogenomic Review Group (IPRG) to review VXDSs, to work on policy development, and, upon request, to advise review divisions on interpretation and evaluation of pharmacogenomic data. Some pharma companies expect advantages from the FDA, thus they favor the VXDSs, while some others are afraid that the competition will use the data for their advantage, and there are examples of this.

The goal is to discover potential markers of longer-term surrogacy, early markers of side effects, markers of stratification, and companion diagnostics.

Often, especially in drug trials that address *me too* compounds, the markers are already available and readily applied.

4.2.8   Early detection and stratification markers

For patient selection, most often precise and validated diagnostics are used (see above). A new trend is based on the concept of early disease detection; this is already widely attempted in cancer treatment, but now being applied to debilitating chronic diseases, such as Alzheimer's disease, osteoarthritis, chronic obstructive pulmonary disease (COPD), obesity, type 2 diabetes, and so forth. The paradigm is aimed at stretching the length of the drug administration, widening the market, and importantly, impacting the treatment outcome via drug prescription to nonsymptomatic patients. Early detection–companion diagnostics are used to identify patients at presymptomatic stages of chronic diseases [15]. There are not many success stories as of yet for early detection; however, a similar approach with multiple FDA approved examples, aimed at increasing efficacy by disease stratification (stratification markers) has been successful though. To this end, many pharma companies are forming alliances and/or purchasing midsize diagnostic companies. The trend is strengthened by a few effective stratification test codevelopments, such as Herceptin for Her2/neu+ positive breast cancer [16], approved in 1998; Gleevec for Philadelphia chromosome positive chronic myeloid leukemia (BCR BCR-ABL ABL) [17], approved in 2001; and the same approved for gastrointestinal stromal tumors (GIST) with C-kit mutations [18], approved in 2003; and Erbitux® for EGFR+ colon cancer [19], approved in 2004. The latest news related thus far is Novartis acquiring Geneoptics a U.S. reference laboratory [20]. This is because the expectation is that the companion diagnostic results will drive drug prescription, thus the incentive is high to control the intellectual property and marketing of these new types of diagnostics.

4.3   Postmarket monitoring

Once drugs are marketed, prescription is often limited to patients with well-characterized conditions. Diagnostic-grade biomarkers are commonly and regularly used to define the conditions, and often include a much larger arsenal of diagnostic procedures than biomolecular diagnostics, such as imaging technologies, various physiological measurements, physical examination, and so forth. Starting in 1998, some drug (Herceptin, Gleevec, Erbitux) prescriptions require specific molecular testing, which defines phenotypically or otherwise undistinguishable conditions as disease stratification markers.

4.3.1   The new trend

The dream of companies developing companion diagnostics–drug combinations is the discovery of highly specific and sensitive robust diagnostics to detect the major survival and quality of life-affecting diseases early in the nonsymptomatic stage. The ultimate expectation is that by identifying the disease early and applying the drug, the disease will not progress beyond the asymptomatic limits, that is, Alzheimer's disease progression will stop before apparent cognitive impairment is detectable, or COPD disease will stop before forced expiratory volume/1st sec of expiration (FEV1) values or vital capacity deviate from the healthy population. An extended version of the theory is that global molecular profiling of all biomolecules (whole-genome sequencing, mRNA profile of blood cells, plasma proteome, and metabolome analysis) will be done as population screening routinely, that is, [21] every half year. Diseases will be detected early and life expectancy will approach 120 years [4]. The paradigm is often questioned because of the challenging task to prove that an early detection marker (a specific marker panel from the global profiling) has sufficient surrogacy value to be indicative of disease progression leading to symptoms or to the lack thereof. In the absence of high-level validation of the surrogacy value at the asymptomatic disease stage, drug diagnostic companies may be accused of *inventing* markers that are simply used to justify prescriptions to an apparently healthy population. Epidemiology and evolution of some of the diseases, for example, Alzheimer's disease is such that a new *drug–IVD* combination would have to be followed in the drug trial context to almost 10 to 15 years, leaving no room to collect revenues on the protected product. Often scientific arguments lean toward the discovery and use of personalized combinations of drugs, which would make testing even more complicated. Despite the issues, many view this new trend as the future solution for the major healthcare issues. The summary of biomarker applications in the drug development process is shown in Figure 4.2a,b.

4.4   Types of biomarkers

To avoid confusion, it has to be clearly stated that the name of the bio-molecular biomarkers is dependent on the application and use. Thus, the same molecular entity can be regarded as an early detection marker or a disease management marker if it provides value in the relevant context. Semantics is also somewhat confusing because the same markers for the same biological applications are often termed differently in the drug development process compared to use in clinical practice. Here, we do not try to clarify semantics, instead when overlapping terminology is possible, both will be mentioned (Figure 4.2a,b).

4.4.1   Disease-specific biomarkers (examples)

These markers may be used for early detection and for disease management depending on the desired application. An example of dual

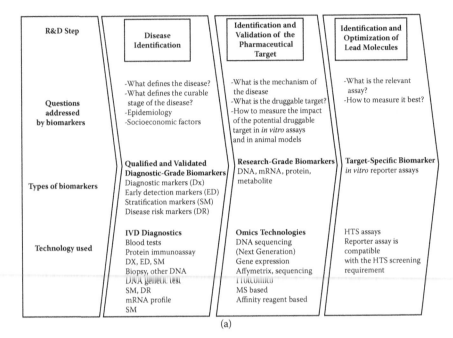

(a)

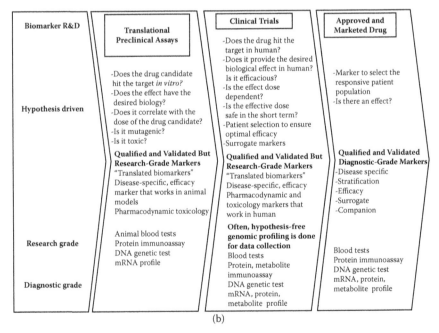

(b)

Figure 4.2 (a) Biomarker applications in the drug R&D process (I). (b) Biomarker applications in the drug R&D process (II).

application is prostate-specific antigen. The need for early detection is the most definitive in the general cancer field, as today the best treatment is still early detection combined with radical surgery and adjuvant chemotherapy. In the examples below, some important trends are described addressing colorectal and prostate cancers [22–26].

The second most frequent cancer worldwide is colorectal cancer (CRC). Its incidence reaches 1 million per year (with a half-million deaths per year). For early diagnosis of CRC, the most popular practice is stool examination for traces of blood. This test, however, is not specific enough (<50%) and it lacks sufficient sensitivity (<80%) so the accuracy is below the necessary level (80–80), therefore there is intensive research to discover more precise and more efficient biomarkers. An epigenetic test, which detects changes in the methylation state of cancer DNA (methylome) from stool samples, seems to be promising and could be useful in the long run. Another promising biomarker set has been discovered by mRNA profiling of blood cells. The studies identified seven possible CRC biomarkers (ANXA3, CLEC4D, LMNB1, PRRG4, TNFAIP6, VNN1, and IL2RB). These show a differential expression profile for individuals with CRC, compared to healthy individuals.

Prostate cancer has also significant mortality in the male population. Prostate-specific antigen (PSA) is a specific protein produced by cancerous prostate cells and to a lower level by normal prostate cells. The use of PSA indicates significant likelihood for the presence of prostate cancer; however, the test is not specific enough (<60%) as some benign diseases (prostate hypertrophy, inflammatory prostate disease, etc.) are also positive. PSA is also useful as a disease management marker, because prostate cancer patients with persistent high PSA levels are less likely to survive. Thus, there is intensive research to find new biomarkers for early detection of prostate cancer. An interesting candidate is careolin-1 (a membrane-associated protein), which has been shown to have greater concentration in plasma samples of prostate cancer patients as compared to a control population. In addition to serum, urine seems to be very useful in the early diagnosis of prostate cancer. A group of scientists found the increased expression of GOLPH 2, SPINK1, and PCA3 transcripts in the urine, which could be possible early-detection biomarkers of prostate cancer.

4.4.2   Prognostic biomarkers (examples)

The prognostic markers provide clinically useful information about the prospects of the patients often before treatment, and often stratify otherwise homogeneous disease categories with respect to treatment outcome, thus the terminology: *stratification* or *prognostic-stratification markers*. As these markers may predict disease or treatment outcomes, they are also referred to as *predictive markers*.

During recent years a few gene expression-based marker panels appeared on the market. Two of these are prototypes of stratification, prognostic-predictive markers, which help in selecting those breast cancer patients who respond to adjuvant chemotherapy [27–34].

Oncotype Dx from Genomic Health is a diagnostic test that quantifies the likelihood of disease recurrence in women with early-stage hormone receptor-positive breast cancer (prognostic significance) and assesses the likely benefit from certain types of chemotherapy (predictive significance). The test measures the expression of 21 genes, and it projects the expected efficiency of therapy, and the likelihood after Tamoxifen™, an adjuvant chemotherapy treatment. Among the 16 predictive genes, the ER, PR, HER-2, and Ki-67 are the most determining ones.

The Mammaprint™ is a DNA assay which detects the expression of 70 genes. It selects the patients, from stage I to IV. The Mammaprint is reported to increase the efficacy of therapy, meanwhile decreasing the costs and side effects.

4.4.3 *Pharmacodynamic biomarkers (examples)*

Pharmacodynamic (PD) biomarkers are markers of a certain pharmacological response, which are of special interest in dose optimization studies. Pharmacodynamic modeling is often based on animal studies and known behavior of certain individual drug classes. The models are used to support initial dosing studies. Two examples include cyclooxygenase inhibitors and a new tyrosine kinase receptor inhibitor [35–38].

The so-called COX (cyclooxygenase) inhibitors are widely applied in the treatment of acute and chronic pain and inflammation, that is, in rheumatoid arthritis and osteoarthritis. The characterization of the correlation between the plasma concentration of COX inhibitors and their pain- and inflammation-decreasing ability is often done via inflammation biomarkers such as PGE2 and thromboxane B2.

SUTENT® by Pfizer, and previously known as *SU11248,* is an oral, small molecule, multitargeted receptor tyrosine kinase (RTK) inhibitor that was approved by the FDA for the treatment of renal cell carcinoma (RCC) and imatinib-resistant gastrointestinal stromal tumor (GIST) on January 26, 2006. Sunitinib was the first cancer drug simultaneously approved for two different indications: in cases of kidney cancer, and for imatinib-resistant or -intolerant gastrointestinal tumor cases. Considering its function, as a tyrosine kinase inhibitor, which selectively inhibits VEGFR-1,-2,-3 RTKs. Soluble VEGF RTKs serve as pharmacodynamic markers for SUTENT therapy.

4.4.4 Efficacy or response (surrogate) biomarkers (examples)

Efficacy biomarkers indicate whether a given treatment achieves the proper results or not, and whether the individual reacts to it or not. Initial attempts to find these markers are the various profiling experiments, which will provide large numbers of correlations with clinical outcome of the treatment, such as treatment efficacy. The candidate correlating markers may qualify in multiple independent clinical trials but often lack any traceable mechanistic relation with the root cause of the disease or the drug target. Surrogate markers derive from candidate correlative candidates; however, real surrogate markers require multiple independent validation cycles and more importantly, acceptance by the medical and regulatory communities. The National Institutes of Health (United States) defines surrogate endpoint as "a biomarker intended to substitute for a clinical endpoint." Primary clinical endpoints are hard to measure (i.e., osteoporosis) and may take an impractically long time to develop (e.g., 10-year survival in the case of the adjuvant chemotherapy) and limit the use. If the surrogate marker is accepted by the medical community and the regulatory agencies, it may be applied instead. However, measurable clinical evidence directly and causatively linked to the primary clinical endpoints are more desirable by regulatory agencies. Replacement of a primary clinical endpoint with a surrogate marker increases the risk of failure in later phase clinical trials, thus a bad decision in choosing the surrogate marker which may reduce the cost of the clinical trial could be the cause of failure of the trial in the long run. Careful judgment is needed and decisions are often based on intuition rather than facts.

Examples for good surrogate markers include an intermediate filament, which is released from apoptotic cancer cells during chemotherapy, and a collagen fragment which indicates bone decay [39–40]. Cyokeratin 18 is an efficient biomarker to measure the effect of breast cancer treatment via combination chemotherapy such as docetaxel or cyclophospha-mide/epirubicin/5-fluorouracil (CEF).

N-terminal telopeptide (or more formally, amino-terminal collagen cross-links, and known by the acronym *NTX*) is a biomarker used to measure the rate of bone turnover. NTX can be measured in the urine or serum of osteoporosis patients, and it can be used to follow treatment efficacy of antiosteoporosis drugs.

4.4.5 Toxicity biomarkers (examples)

Sensitive detection of toxic effect is of key importance. Toxicity bio-markers report effects that precede significant tissue damage and organ failure. The classic examples include detection of liver enzymes in the blood.

Table 4.1 The Most Important Applications for Individual Biomolecular Biomarker Classes

Type of Biomarkers	Drug Development	Disease Management
Disease specific	—	IVD diagnostic
Prognostic	—	Treatment planning and responder selection
Predictive	Selection of responding patient population	Selection of responding patient population
Pharmacodynamic	It proves that the medicine reaches the target and that the effective dose is achieved	—
Efficacy	It provides clues for the patient's reaction, before it may be evident (surrogate endpoint)	—
Toxicity	Early detection of toxicity to avoid organ and tissue damages	To remove those patients from the treatment who are oversensitive

The liver plays a central role in transforming and clearing chemicals and is susceptible to the toxicity from these agents. Certain medicinal agents when overdosed and sometimes even when introduced within therapeutic ranges may injure the organ. More than 900 drugs have been implicated in causing liver injury and it is the most common reason for a drug to be withdrawn from the market. Chemicals often cause subclinical injury to liver, which manifests only as abnormal liver enzyme tests. Drug-induced liver injury is responsible for 5% of all hospital admissions and 50% of all acute liver failures.

A group of enzymes located in the endoplasmic reticulum, known as *cytochrome P-450*, is the most important family of metabolizing enzymes in the liver. Cytochrome P-450 is the terminal oxidase component of an electron transport chain. It is not a single enzyme, but rather consists of a closely related family of 50 isoforms; six of them metabolize 90% of drugs. There is a tremendous diversity of individual P-450 gene products, and this heterogeneity allows the liver to perform oxidation on a vast array of chemicals (including almost all drugs). The most frequently used biomarkers of liver cell damage are alanin transferase and alkaline phosphatase. [41,42].

A simplified table (Table 4.1) shows the most important application for each biomolecular biomarker class.

References

1. Bloom, J. C. and Dean, R. A. (eds.). 2003. *Biomarkers in Clinical Development.* New York: Marcel Dekker.
2. Kuhn, T. 1965. *The Structure of Scientific Revolutions.* Chicago: University of Chicago Press.
3. Drmanac, R. 2011. The advent of personal genome sequencing. *Genet. Med.* 13(3):188–90.
4. Hempel, W., Sziraczky, G., Swalm, L., and Takacs, L. 2008. Analytical Paper: Biomarkers: Impact on Biomedical Research and Healthcare: Case Reports Directorate for Science and Technology and Industry Committee for Science and Technological Policy. Biosystem International, Evry, France and Armus Corporation, California, United States. http://www.oecd.org/dataoecd/48/3/41892771.pdf.
5. Ahmed, F. E. 2009. Utility of mass spectrometry for proteome analysis: Part II. Ion-activation methods, statistics, bioinformatics and annotation. *Expert Rev. Proteomics* 6(2):171–97.
6. Ellington, A. A., Kullo, I. J., Bailey, K. R., and Klee, G. G. 2010. Antibody-based protein multiplex platforms: Technical and operational challenges. *Clin. Chem.* 56(2):186–93.
7. Ressom, H. W., Varghese, R. S., Zhang, Z., Xuan, J., and Clarke, R. 2008. Classification algorithms for phenotype prediction in genomics and proteomics. *Front Biosci.* 13:691–708.
8. Sweep, F. C., Fritsche, H. A., Gion, M., Klee, G. G., and Schmitt, M. 2003. EORTC-NCI Working Group. Considerations on development, validation, application, and quality control of immune (metric) biomarker assays in clinical cancer research: An EORTC-NCI working group report. *Int. J. Oncol.* 23(6):1715–26.
9. Rodriguez, H., Tezak, Z., Mesri, M., Carr, S. A., Liebler, D. C., Fisher, S. J., et al. 2010. Workshop Participants. Analytical validation of protein-based multiplex assays: A workshop report by the NCI-FDA interagency oncology task force on molecular diagnostics. *Clin. Chem.* 56(2):237–43.
10. Silverman, L., Campbell, R., and Broach, J. R. 1998. New assay technologies for high-throughput screening. *Curr. Opin. Chem. Biol.* 2:397–403.
11. Campbell, J. B. 2010. Improving lead generation success through integrated methods: Transcending "drug discovery by numbers." *IDrugs* 13:874–9.
12. Nakao, K., Yasoda, A., Ebihara, K., Hosoda, K., and Mukoyama, M. 2009. Translational research of novel hormones: Lessons from animal models and rare human diseases for common human diseases. *J. Mol. Med.* 87(10):1029–39.
13. Ray, K. K. and Cannon, C. P. 2005. Atorvastatin and cardiovascular protection: A review and comparison of recent clinical trials. *Expert. Opin. Pharmacother.* 6:915–27.
14. FDA. 2005. Processing and Reviewing Voluntary Genomic Data Submissions (VGDSs): http://www.fda.gov/downloads/AboutFDA/CentersOffices/CDER/ManualofPoliciesProcedures/ucm073575.pdf.
15. Dunn, B. K., Wagner, P. D., Anderson, D., and Greenwald, P. 2010. Molecular markers for early detection. *Semin. Oncol.* 37:224–42.
16. Dean-Colomb, W. and Esteva, F. J. 2008. Her2-positive breast cancer: Herceptin and beyond. *Eur. J. Cancer* 44(18):2806–12.

17. Nadal, E. and Olavarria, E. 2004. Imatinib mesylate (Gleevec/Glivec), a molecular-targeted therapy for chronic myeloid leukaemia and other malignancies. *Int. J. Clin. Pract.* 58(5):511–6.

18. Zoubir, M., Tursz, T., Ménard, C., Zitvogel, L., and Chaput, N. 2010. Imatinib mesylate (Gleevec): Targeted therapy against cancer with immune properties. *Endocr. Metab. Immune Disord. Drug Targets* 1:10(1):1–7.

19. Bou-Assaly, W. and Mukherji, S. 2010. Cetuximab (erbitux). *AJNR Am. J. Neuroradiol.* 31(4):626–7.

20. Novartis. 2011. Novartis Announces Agreement To Acquire Genoptix, Inc. in All Cash Offer: http://www.novartis.com/newsroom/media-releases/en/2011/1481651.shtml.

21. Hood, L. and Friend, S. H. 2011. Predictive, personalized, preventive, participatory (P4) cancer medicine. *Nat. Rev. Clin. Oncol.* 8(3):184–7.

22. Marshall, K. W., Mohr, S., El Khettabi, F., Nossova, N., Chao, S., Bao, W., Ma, J., Li, X. J., and Liew, C. C. 2010. Blood-based biomarker panel for stratifying current risk for colorectal cancer. *Int. J. Cancer* 126:1177–186.

23. Yip, K. T., Das, P. K., Suria, D., Lim, Ch. R., Ng, G. H., and Liew, C. C. 2010. A case-controlled validation study of a blood based seven-gene biomarker panel for colorectal cancer in Malaysia. *J. Exp. Clin. Canc. Res.* 29:128.

24. Thompson, T. C., Tahir, S. A., Li, L., Watanabe, M., Naruishi, K., Yang, G., Kadmon, D., Logothetis, C. J., Troncoso, P., Ren, C., Goltsov, A., and Park, S. 2010. The role of caveolin-1 in prostate cancer: Clinical implications. *Prostate Cancer P. D.* 13(1): 6–11.

25. Laxman, B., Morris, D. S., Yu, J., Siddiqui, J., Cao, J., Mehra, R., Lonigro, R. J., Tsodikov, A., Wei, J. T., Tomlins, S. A., and Chinnaiyan, A. M. 2008. A first-generation multiplex biomarker analysis of urine for the early detection of prostate cancer. *Cancer Res.* 68(3):645–9.

26. Dudley, J. T. and Atul, J. B. 2009. Identification of discriminating biomarkers for human disease using integrative network biology. *Pacific Symposium on Biocomputing* 14:27–38.

27. Bao, T. and Davidson, N. E. 2008. Gene expression profiling of breast cancer. *Adv. Surg.* 2042:249–60.

28. Glas, A. M., Floore, A., and Delahaye, L. J. 2006. Converting a breast cancer microarray signature into a high-throughput diagnostic test. *BMC Genomics* 7:278.

29. Weigelt, B., Hu, Z., and He, X., 2005. Molecular portraits and 70-gene prognosis signature are preserved throughout the metastatic process of breast cancer. *Cancer Res.* 65:9155–58.

30. Ross, J. S., Hatzis, C., Symmans, W. F., Pusztai, L., and Hortobágyi, G. N. 2008. Commercialized multigene predictors of clinical outcome for breast cancer. *The Oncologist* 13:477–93.

31. Paik, S., Tang, G., and Shak, S. 2006. Gene expression and benefit of chemotherapy in women with node-negative, estrogen receptor-positive breast cancer. *J. Clin. Oncol.* 24:3726–34.

32. Paik, S. 2007. Development and clinical utility of a 21-gene recurrence score prognostic assay in patients with early breast cancer treated with tamoxifen. *The Oncologist* 12:631–35.

33. Flanagan, M. B., Dabbs, D. J., Brufsky, A. M., Beriwal, S., and Bhargava, R. 2008. Histopathologic variables predict Oncotype Dx™ recurrence score. *Mod. Pathol.* 21:1255–61.

34. Oakman, C., Bessi, S., Zafarana, E., Galardi, F., Biganzoli, L., and Leo, A. D. 2009. New diagnostics and biological predictors of outcome in early breast cancer. *Breast Cancer Res.* 11:205.
35. Huntjens, D. R. H., Danhof, M., and Della Pasqua, O. E. 2005. Pharmacokinetic–pharmacodynamic correlations and biomarkers in the development of COX-2 inhibitors. *Rheumatology* 44:846–59.
36. DePrimo, S. E., Bello, C. L., Smeraglia, J., Baum, C. M., Spinella, D., Rini, B. I., Michaelson, M. D., and Motzer, R. J. 2007. Circulating protein biomarkers of pharmacodynamic activity of sunitinib in patients with metastatic renal cell carcinoma: Modulation of VEGF and VEGF-related proteins. *J. Transl. Med.* 5:32.
37. Straubinger, R. M., Krzyzanski, W., Francoforte, C. M., and Qu, J. 2007. Applications of quantitative pharmacodynamic effect markers in drug target identification and therapy development. *Anticancer Res.* 27:1237–46.
38. Workman, P., Aboagye, E. O., Chung, Y. L., Griffiths, J. R., Hart, R., Leach, M. O., et al. 2006. Minimally invasive pharmacokinetic and pharmacodynamic technologies in hypothesis-testing clinical trials of innovative therapies. *J. Natl. Cancer. Inst.* 98:580–98.
39. Olofsson, M. H., Ueno, T., Pan, Y., Xu, R., Cai, F., Kuip, H., Muerdter, T. E., et al. 2007. Cytokeratin-18 is a useful serum biomarker for early determination of response of breast carcinomas to chemotherapy. *Clin. Cancer Res.* 13:11.
40. Abe, Y., Ishikawa, H., and Fukao, A. 2008. Higher efficacy of urinary bone resorption marker measurements in assessing response to treatment for osteoporosis in postmenopausal women. *Exp. Med.* 214:51–59.
41. Vaidya, V. S., Ozer, J. S., Dieterle, F., Collings, F. B., Ramirez, V., Troth, S., et al. 2010. Kidney injury molecule-1 outperforms traditional biomarkers of kidney injury in preclinical biomarker qualification studies. *Nat. Biotechnol.* 28(5):478.
42. Miyamoto, M., Yanai, M., Ookubo, S., Awasaki, N., Takami, K., and Imai, R. 2008. Detection of cell-free, liver-specific mRNAs in peripheral blood from rats with hepatotoxicity: A potential toxicological biomarker for safety evaluation. *Toxicol. Sci.* 106(2):538–45.

chapter five

Quo vadis *biomedical sciences in the omics era*

Toward computational biology and medicine

András Falus,[*] Éva Pállinger, Gergely Tölgyesi, Viktor Molnar, Tamás Szabó, Edit Buzás, Katalin Éder, Peter Antal, and Csaba Szalai

Contents

[*] Corresponding author.

5.1   Systems and computational biology and medicine

Rapid improvements in accessibility to molecular databases as well as availability of high-throughput genomic, proteomic, and other omics methodologies are forcing a considerable shift in research and development strategies for biomedicine. The recent change in the research paradigm focusing on biology as a sort of system-science is still difficult to grasp. *Systems biology*, a kind of holistic science exploring complex interactions in biological systems, is currently closely related to the development and application of bioinformatic and biostatistical tools trying to manage vast quantities of genomic and proteomic data.

Clinical sciences translate achievements of basic research to medically relevant applications, that is, diagnosis, prediction, prevention, and therapy. It covers a broad area of complex disorders of the human system including proliferative disorders (cancer and degenerative diseases), cardiovascular syndromes, various inherited and acquired deficiencies, autoimmunity, allergy, physiological and pathological responses to pathogens and transplantation, and various therapeutical approaches [1,2].

The exceptional complexity of the human biological (including immunological, neuronal, endocrine, circulation, etc.) system has several sources: combinatorial variability, plasticity, degeneracy, and adaptivity of the biological system; interactions of the immune system and other self cells, tissues, and organs; the variability of pathogens, self and environmental antigens; and the maintenance of multiple regulatory pathways. There are incredible values based on *magic, cosmic numbers* of human biology (3.2×10^9 nucleotides/haploid human genome, 2 m DNA in each of the 10^{14} cells of a human body, representing over 5 light days!!). In one human body, 100^{23} possible protein molecules only for the average size of 100 amino acids represent only the top of the complexity *iceberg*. The structural and dynamic aspects of combinatory figures of possible networks both at gene and protein levels are enormous. This complexity ensures that huge amounts of data must be produced and analyzed for deciphering the workings of the living system. In order to use these huge data banks effectively enough, we need increasingly sophisticated tools of biostatics, bioinformatics, and mathematical modeling; in other words, the principal elements of computational biology and medicine. The integrative approaches combining multiple tools are particularly important in clinical research, comprising large cohorts of usually nonhomogenous groups (i.e., *endophenotypes*) where disease phenotype, clinical progression, molecular profiles, and patient characteristics show huge variation.

Medical sciences are not clearly differentiated as a clinical specialty because they involve a number of medical disciplines.

5.1.1 *Expansion of biomedical knowledge*

A major driver for the advancement of clinical science is the dramatic expansion and accessibility of knowledge of structural and functional elements of the living system at the biochemical, cellular, organ, organism, and population levels. This knowledge is largely accumulated thanks to scientific and technological progress, which continues unabated; the volume of scientific information is estimated to double every 15 years. The source of key enabling technologies is genomics [3], proteomics, and metabolomics [4], as well as bioinformatics and computational biology [5] (including genomic pathway and gene set enrichment analysis). Since these high-throughput techniques provide an extremely large mass of data, an entirely novel dimension of information seems to be generated for describing molecular profiles of various physiological and pathological states. The rapid development of new generation sequencing technologies seems to play a basic platform for the expected remodeling of omics technologies. This very similar *rearrangement* of wet-lab instrumentation is obviously motivated by the decrease in the expenses of large-scale sequencing. Advanced methods for quantification of biological responses provide solid toolsets for detailed study of human pathology and a complex set of environmental interactions.

Flow cytometry enables measurement and characterization of gene/protein expression of individual cells and molecules representing various experimental states. Technical improvement of assays for monitoring (e.g., multiparametric flow cytometry, nanotechnology for quantization of cytokine production, enzyme-linked immunoassay on spot (ELISPOT), intracytoplasmic cytokine staining, and mRNA as well as microRNA-based assays) continuously expands our ability to measure profiles of bioactive molecules (such as cytokines or lipid mediators) that direct and modulate biological responses in health and disease [6]. Latest developments in laser scanning cytometry allow the measurement and analysis of effector function of individual cells *in situ*, thus representing molecular and cellular events in physiological and pathological states [7].

This volume, exemplifed by various topics in clinical research, introduces various tools of genomics as well as those of computational biology. This collection of topics is not a complete collection of works in this field.

Rather, it is a starting point where examples of various *wet lab* and computational approaches are studied for advancement of knowledge and translation of these results into new products, methods, and therapies. Translating basic research advances into medical applications increasingly requires a multidisciplinary approach and teams comprising clinicians at the bedside, basic scientists at the laboratory bench, engineers who develop advanced instrumentation, along with biostatisticians and bioinformaticians who perform data analyses and interpretations.

5.1.2   Computational biology—Biosystemomics

Computational biology—a kind of *biosystemomics*, therefore, is a powerful new technology that combines basic and clinical research with high-throughput instrumentation and bioinformatics for the analysis and interpretation of the data. Biosystemomics is similar to genomics and proteomics in that a major challenge is to understand the web of genes and proteins involved in the functioning of the biological system. In addition, biosystemomics must address factors arising from the complex micro- and macroenvironment affecting the function, as well as external challenges arising from environmental diversity and resulting in epigenetic differences. Systems biology and medicine screening of markers of the biological status will in the near future be used to determine who should be enrolled in or excluded from a particular clinical trial and follow-up of these markers throughout the course of therapy [8,9]. In addition, biosystemomics approaches herald the development of the new generation of tools for prediction, prevention, and therapies to be tailored precisely to both the genetic makeup of the human population and of the disease profile, be it cancer, allergy, or infection [10,11].

This is an insight into systems biology, in which all main components and their interactions within a biological system are measured, and then assembled into modules for further study. This approach makes no assumptions about underlying mechanisms—the measurement is direct; the disadvantage is that the scale of information that can be obtained is enormous. Systems biology uncovers relations between entities in a biological system and their regulation.

Systems biology provides integration of the approaches of molecular-cellular biology, bioinformatics, genomics, proteomics, network (graph theorem), and other related scientific fields with the aim to derive integrated models of medical modulatory processes. This chapter focuses on the newest trends of system-based approaches to characterize in detail the molecular mechanisms of regulatory processes in biological responses.

5.2 Present trends and future perspectives in major omics technologies

5.2.1 Flow cytometry in the omics era: A central segment of proteomics

There is a new attitude in medicine these days: personalized medicine, based on molecular diagnostics, comes into focus. This practice could not have evolved without new high-throughput screening (HTS) laboratory methods, which can detect many analytes in a very short time. On the other hand, the analysis of so many results created by HTS techniques could not be explained without the gigantic development of bioinformatics [12].

The conception of personalized medicine has two trends: it needs the complex knowledge of the pathomechanisms of biological processes (*top-down*), and aims to study and validate already known and potential biological networks (*bottom-up*).

The application of the term *proteomics* [13,14], and the development of new HTS laboratory techniques (e.g., immunoanalytical methods, flow cytometry, microarray technologies, and mass spectrometry) open new opportunities in diagnostics. It should be noted that during proteome analysis, (1) there are many differences between the protein content and the protein pattern of different cell populations and (2) many factors (e.g., environmental effects, time, activation, and differentiation states) can cause detectable changes in the proteome of an individual cell.

Molecular cell biology, the so-called cytomics, summarizes and explains the structural and functional properties of proteins and their expression. The aim of cytomics, widespread functional cell analysis, are the following: (1) the complex mapping of the molecular mechanisms of biological processes, (2) the discovery of new biomarkers, (3) the selection of new therapeutic targets, (4) the reformation of disease classification, (5) the expansion of modern therapy monitoring, and (6) the finding of new predictive and prognostic markers.

One of the most powerful technologies for the analysis of cytomics is flow cytometry. Multiparametric flow cytometry can serve as a bridge between proteomics and cytomics, since one is capable of characterizing the functional and morphological properties of single cells at the same time with this method.

Flow cytometry has undergone a great transformation during its five-decade history; however, it has been a HTS technology right from the beginning. The main application of the early instruments was the characterization of circulating leukocytes. These cytometers could measure three different parameters (two scattered light signals and one fluorescence signal). By contrast, many new applications have been adapted into modern, multicolor flow cytometers. A widely applied technology is the microsphere-based flow cytometric assay, which is usable for the detection of soluble molecules in biological fluids. Currently, it seems not to be utopist to investigate protein–protein interactions by fluorescence resonance energy transfer (FRET). With the development of monoclonal antibodies specific to modified (e.g., phosphorylated) proteins, measurement of signal transduction pathways of single cells has also become available. Considering that more and more fluorescent dyes can be detected simultaneously, immunophenotyping can be combined with intracellular staining, which means proteomics at the single-cell level.

5.2.1.1 Multicolor measurements—Polychromatic flow cytometry [15–17]

It was necessary to increase the number of simultaneous detectable fluorescent dyes, in order to develop the investigation of complex cellular networks in the background of physiological and pathophysiological processes. Polychromatic flow cytometry—measurement of six or more fluorochromes at the same time—can be a key tool for the understanding of cellular and protein networks. Using more colors in the same experiment permits even more sensitive and more precise measurements of rare cell subsets as well. This benefit is important not only in vaccine development, but also in other fields of immunology (antigen-specific T cell measurements; determination of T cell subpopulations by their cytokine production or by their protein pattern; follow-up of HIV infection, etc.), hematology (classification of hematological malignancies, detection of minimal residual diseases, etc.), and oncology (characterization of antitumor immunity, e.g., investigation of tumor–antigen-specific T cells).

Another important advantage of polychromatic flow cytometry is the measurement of rare cell populations. Multichannel detections allow applying the so-called dump channel. Undesired cells can be excluded from analysis using this dump channel after their staining by specific antibodies, labeled with the same fluorochromes. Using the other fluorescent channels (dyes), low-frequency cell subsets can be detected in a very sensitive manner.

5.2.1.2 Fluorescence-activated cell sorting

Besides multicolor immunophenotyping, flow cytometers display a unique function: they are the only instruments capable of separating a heterogeneous suspension of cells into purified fractions (*sorting*) on the basis of light scattering and fluorescence properties. It has to be emphasized that fluorescence-activated cell sorting is the only technology that can separate single cell types with high speed and purity. During the past few years, cell sorting has undergone several changes and the trend is clear: the traditional low-speed instruments (approximately 1000 cells/sec) have been replaced by the generation of high-speed cell sorters (more than 100,000 cells/sec) [18]. The significance of high-speed cell sorters can be grouped into two categories: (1) Scientists use *bulk sorting* when further experiments need a large amount of cells and the sample contains a high percentage of subsets expressing a given phenotype. (2) Rare event sorting has to be used when the concentration of the investigated cell types is less than 0, 1% in the sample. In these cases significantly higher amount of cells have to be analyzed.

5.2.1.3   *Multiplexed microsphere-based flow cytometric assay [19]*
Although the majority of flow cytometric applications are cell based, smaller particles (microbes or synthetic microspheres) can also be detected. Conventional fluorescence-activated cell sorter (FACS) machines with appropriate optical configuration can acquire microsphere-based light scattering and emission data. Proteins, oligonucleotides, peptides, lipids, or carbohydrates can bind to the surface of synthetic particles, in order to catch different analytes and can be visualized by fluorochrome-conjugated detection molecules. Although the most current applications are cytokine quantization [20], single nucleotide polymorphism (SNP) (see more in Sections 2.2 and 2.4 in Chapter 2), genotyping [21], and phospho-protein analysis [22], other immunoassays are also available. Moreover, the prospects of the method are unlimited. Microsphere-based technology allows (1) detection of multiple analytes in biological and environmental fluids, (2) minimalization of sample volume, (3) reduction of cost per test, and (4) preservation of time.

5.2.1.4   *Phospho-specific flow cytometry [22,23]*
The study of complex interactions in biological systems is essential in the new omics era.

Flow cytometry is the connection between single cell measurements and multiple protein analysis. Immunophenotyping combined with phospho-specific protein analysis provides information about cell functionality, modulated by environmental effects (single cell-based analysis). On the other hand, multiplexed microsphere-based assays allow ultrasensitive high-throughput protein measurements. Both applications are very important in drug discovery, both in the identification of new targets and in the follow-up of drug effects.

5.2.2   *High-throughput genotyping and the genome-wide association studies*

The omics era provides an exceptional possibility to approach the molecular cell functions simultaneously at multiple levels (Figure 5.1), providing

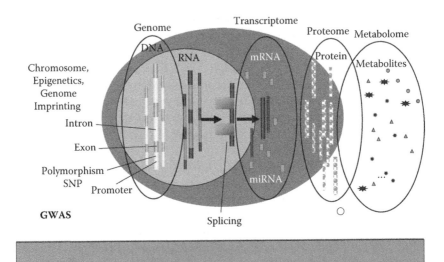

Figure 5.1 Approaches of the biological complexity in the *omics* era.

a huge amount of experimental data from DNA-RNA-protein metabolomic levels.

Genome-wide association studies (GWAS) compare the distribution of single nucleotide polymorphisms (SNPs) (see more in Section 3.4 in Chapter 3), and recently also copy number variations (CNVs)—using arrays that can examine 100,000 to >1.8 million markers at a time—in hundreds or even thousands of people with and without a particular phenotype. Presently this technique (high-throughput genotyping) is one of the most powerful and most frequently published methods in population genomics, which revolutionized the association studies revealing thousands of genetic variants (mainly SNPs) associating with a given phenotype. An indication of its significance is that it was selected the "Breakthrough of the Year 2007" by one of the most distinguished scientific journals, *Science* [24]. Since then, a Web page has been established named *A Catalog of Published Genome-Wide Association Studies* [25]. In this Web page, those publications are listed which assayed at least 100,000 SNPs in the initial stage. Only 2.5 years after its initiation, more than 600 publications and more than 3000 SNPs have been cataloged on this page. One can ask why this method is so popular and what we can gain from its results?

It is well known that practically all of our traits have some genetic background. For instance, susceptibility to a disease is influenced by our outer or inner features; how we react to a drug or therapy, are all influenced by our genetic background. Before the genomic era, all of these were studied by candidate gene association studies, in which, according to its known function, a gene was selected and investigated as to whether its variants influenced a given phenotype. This approach revealed some important genetic polymorphisms that play a role in certain phenotypes, but as our knowledge increased, it turned out that most traits are influenced by several hundred or even several thousand factors, including genetic and environmental ones. The actual phenotype of an individual is the result of the interaction of these factors. This could not be properly studied by investigating one or two factors at a time. In the meantime, together with the genomic area, new methods have been developed with which more and more variants could be genotyped with one measurement. Now, the top methods with >1.8 million probes can analyze the majority of the frequent SNPs and haplotypes of the whole genome, so theoretically, it became possible to solve the puzzle of genetic backgrounds. And indeed, with very strict statistical methods and study design, and including in some studies several tens of thousands of participants, a huge amount of interesting phenotype-influencing variants have been revealed [26]. For example, in type 1 diabetes mellitus, more than 260 genes were found to influence the susceptibility to the disease, or 50 common variants were detected influencing the height [27]. Huge consortiums were founded like the Wellcome Trust Case Control Consortium (WTCCC), whose aims were to exploit progress in understanding the patterns of human genome sequence variation, along with advances in high-throughput genotyping technologies, and to explore the utility, design, and analyses of GWAS. In this consortium, 19,000 samples were collected. In WTCCC2, initiated in 2008, studies on as many as 120,000 samples are planned [28].

However, it turned out that something was wrong with these methods. For example, in height, which has a heritability of ~80%, only approximately 5% has been explained by the 50 common polymorphisms identified by GWAS. It was similar in other cases as well. There are several explanations for this. One of the theories states that these common traits are influenced by rare variants rather than by common ones, and rare variants cannot be measured by this method [29]. The other popular theory states that the traditional statistical methods are not suitable for analyzing so many variants at the same time, and are unable to find the interactions, which can be responsible for a considerable fraction of the heritability.

Probably, there is truth in both explanations. For the statistical problem, new methods are in development. A solution might be offered by the Bayesian statistical framework providing inherent correction for model complexity (i.e., multiple testing) and model averaging for finding significant aggregated hypotheses. This framework can be applied for both conditional models such as logistic regression, and domain models such as Bayesian networks. It allows the analysis of associations at pairwise level, and multivariate and interaction levels, which provides a new, unified, interlinked, and multilevel view in gene association studies with superior performance (for details see Section 4.2 in Chapter 4).

How can we utilize the findings of the GWAS?

Several companies have already been developed for personal genomics, exploiting the results. Because the method is quite cheap, some offer for a few hundred dollars a complete genotyping with some evaluation including disease susceptibilities, ancestry and family tree testing, or how one will react to certain therapies. Although, because of the earlier mentioned problems, the predictions of these companies are still controversial, it is probable that as our knowledge of genomics and bioinformatics increases, the predictions will be more and more precise. Nevertheless, it must be added that, because of the unpredictable environmental factors and also because of the sometimes stochastic nature of the genomic interactions, these predictions may never be certainties, only probabilities.

On the other hand, all genes and metabolic pathways revealed by GWAS are potential therapeutic and diagnostic targets. Even today, several pharmacological companies are working and have been established to utilize these results, searching for novel diagnoses, drugs, or therapies. The results of GWAS may also be used for personal therapies. Since the measurement is quite inexpensive and simple, it can be hypothesized that in the future, this will become part of the routine protocol, and each individual can have a personal file, which contains information, among others, about his/her genomic data. All of these data can be put on a portable device (e.g., personal chip, smart phone, pen drive) and can be used, for example, by physicians for decision support. Technically, we already have almost all the necessary infrastructure for this, except for the proper bioinformatic methods, and it requires much more data and experience with genotype–phenotype relationships. This data gathering and evaluation is in progress, but it needs 10 or more years before widespread use (if ever). In addition, there are also several sociological and ethical questions in connection with this type of personal genomics to be coped with, which might need even more time.

5.2.3 Next-generation sequencing

DNA sequencing means a reading of the nucleotide order of a piece of DNA. The main purpose of the Human Genome Project (HGP) was to determine the nucleotide composition of the human genome. This was accomplished with an immense international effort for $3 billion between 1990 and 2003 [30]. It also became apparent how little we knew about the human genome and how valuable it would be if a human genome had been sequenced for a much cheaper price and less time. This initiated a huge competition for inexpensive and quicker methods. This competition has brought about quite unexpected and quick success. In 2007, next-generation sequencing (NGS) was selected as *the method of the year* and since then the development has continued with similar speed [31]. Interestingly, this next-generation sequencing is not one method, but several companies developing different revolutionary techniques in parallel. In 2010, there were companies which offered DNA-sequencing equipment and methods with which a human genome could be sequenced for $10,000 in 7 days (compared to $3 billion and 13 years of HGP), and almost every month companies announce more inexpensive, quicker, or more exact sequencing. Now it seems that we are not far away from the possibility that the genome of every individual can be sequenced for a relatively low price. But, this is only one side of the coin. Similar to the problems mentioned in the case of high-throughput genotyping, the bottleneck here is also bioinformatics. The human genome consists of 3.2 billion nucleotides. For good quality data with the present methods, a genome must be sequenced at least 40-fold to reduce the amount of sequencing errors, and because the sequence should be put together from small pieces, the whole process generates a huge amount of data [32]. Some sequencing instruments can generate terabytes of data a day. To store and evaluate this amount of data is an immense challenge for the informatics.

What is the advantage and disadvantage of NGS compared to high-throughput genotyping (HTG)? Naturally, as NGS determines all nucleotides to be sequenced, it can detect rare and even novel variants, while by HGT, only common and known variations can be measured. Thus, there is a possibility of determining the missing heritable fraction of common traits for which the rare variants are responsible. It will also solve the problems with monogenic disorders. It can also have significance, for example, in tumor diagnosis. Sequencing both the healthy and the tumor genome in an individual could help in the exact classification of the tumor, offering new information about tumor pathogenesis, and might help in tailoring personal and more effective therapies.

On the other hand, each human genome contains a couple of million SNPs and thousands of unknown or novel variants (new alleles constantly arise, at an estimated rate of approximately 175 per diploid human genome per generation). Besides the huge number to cope with, there are several difficulties with the results. First, real mutations must be distinguished from the sequencing errors, and second, it should be estimated what the effect of a mutation is; whether it is functional or not? As our knowledge about the human genome is still very restricted, this last one is in most cases almost impossible, and requires a lot of time-consuming and difficult, *wet* (real, not only *in silico*) laboratory experiments.

The difficulties associated with NGS are even more serious than in the case of HTG, but naturally, we can also gain more information from it. All of the challenges associated with the storage, handling, and evaluation of the nucleotide sequence of the human genome are immense, requiring time and a lot of money to solve. However, consortiums, universities, (bio) informaticians, research, and profit-oriented institutions work hard on these problems, and if the problems are solvable, they will be solved in the next years.

How can we utilize the results of NGS in the future? Everything discussed in the last two paragraphs of the HTG part is even truer for NGS. As for which of these methods will play a greater role in our everyday life in the future—we do not know yet, but hopefully, we will benefit from it.

5.2.4   Genotyping: Comparative genome hybridization, copy number variants, and single nucleotide polymorphism

The genome itself is a fruitful area for exploration, and two technologies in particular are having major impacts on biomedical studies: comparative genome hybridization (CGH) and single nucleotide polymorphism (SNP) genotyping. SNPs are the predominant source of variation in the human genome: it is estimated that any two individuals (other than identical twins) have at least 6 million individual base differences in their genome sequence. It is consequently unsurprising that a growing focus on medical and population genetics studies are genome-wide association studies for identifying regions of the genome that may harbor genes responsible for complex phenotypes. The ability to carry out such studies has been revolutionized by the development of microarray platforms (see details in Section 2.5 in Chapter 2) that facilitate very-high-throughput and cost-effective genotyping of hundreds of thousands to millions of SNPs on a single array. There are two major microarray platforms commonly used for whole-genome association studies, Affymetrix and Illumina, which fundamentally differ in their SNP detection assays. Affymetrix arrays detect polymorphisms by a hybridization assay, whereas the Illumina platform relies upon enzymatic extension of genomic DNA captured at specific sequences on bead arrays.

It has been well known that genomic deletions and duplications are important in human disease and are often associated with various cancers, unexplained developmental delay/intellectual disability, autism spectrum disorders, and multiple congenital anomalies. In addition, it is becoming increasingly apparent that variations in DNA copy number make a major contribution to genetic variation in human populations. Recent studies of apparently healthy individuals have found that individual variation comes not only from well-known SNPs, but also from structural variants involving large segments of DNA, so-called copy number variants (CNVs). CNVs are defined as structural variations in the genome (submicroscopic variants, including deletions, insertions, duplications, and large-scale copy number variants typically at least 1 kb in size) for which copy number differences have been observed between two or more genomes. CNVs can vary from a simple segmental duplication to a highly complex genomic rearrangement involving multiple elements. The largest database that has catalogued CNVs is the Database of Genomic Variants [33]. The number of CNVs identified has increased dramatically as the resolution of detection technologies has improved. Consequently, the ability to examine genomic DNA for alterations to the normal diploid condition is highly desirable. Array-CGH, as the name implies, is a technique for comparing genomic DNA via microarray hybridization. DNA is prepared from a normal reference source and a sample of interest, labeled with different fluorescent dyes, and hybridized to an appropriate array. Increases in copy number result in higher signals in the sample channel and deletions yield higher signals in the control channel. The log ratios for each probe are plotted in order across the genome with deletions showing a positive ratio and amplifications a negative ratio. The first whole-genome array-CGH spotted arrays contained bacterial artificial chromosome (BAC) clones covering the entire genome at a relatively low resolution (150 kb to 1 Mb). In spite of their low resolution, BAC arrays are still popular due to their robustness, especially in prenatal array-CGH applications, where in many cases the test sample DNA obtained from fetal amnion or chorionic villi cells has relatively low quality and quantity. In the past years, the *in situ* synthesis methods permit the fabrication of arrays containing millions of oligonucleotide probes and allow tiling of the human genome at densities approaching a 50- to 60-mer probe every kb, which drastically increased the diagnostic yield of the clinical genetic testing of individuals with unexplained developmental delay/intellectual disability (DD/ID), autism spectrum disorders (ASD), or multiple congenital anomalies (MCA). Microarray-based genomic copy number analysis is now a commonly ordered clinical genetic test for this patient population and is offered under various names, such as multiple congenital anomalies (MCA) and *molecular karyotyping*. The classical G-banded karyotype analysis allows a cytogeneticist to visualize and analyze chromosomes for chromosomal rearrangements, including genomic gains and losses. CMA performs a similar function, but at a much higher resolution for genomic imbalances. Determining the clinical significance of variants identified by CMA can be challenging. Although CMA offers the sensitivity of high-resolution genome-wide detection of clinically significant CNVs, the additional challenge of interpreting variants of uncertain clinical significance (VOUS) can impose a burden on clinicians and laboratories. To solve this issue, an international consortium of world leading genetics experts has issued a consensus statement about the use of *chromosomal microarray* testing. The International Standard Cytogenomic Array (ISCA) Consortium [34] conducted a literature review of 33 studies, including 21,698 patients tested by CMA. The ISCA provided an evidence-based summary of clinical cytogenetic testing comparing CMA to G-banded karyotyping with respect to technical advantages and limitations, diagnostic yield for various types of chromosomal aberrations, and issues that affect test interpretation. CMA offers a much higher diagnostic yield (15% to 20%) for genetic testing of individuals with unexplained DD/ID, ASD, or MCA than a G-banded karyotype (~3%, excluding Down syndrome and other recognizable chromosomal syndromes), primarily because of its higher sensitivity for submicroscopic deletions and duplications. Truly balanced rearrangements and low-level mosaicism are generally not detectable by arrays, but these are relatively infrequent causes of abnormal phenotypes in this population (<1%). Available evidence strongly supports the use of CMA in place of G-banded karyotyping as the first-tier cytogenetic diagnostic test for patients with DD/ID, ASD, or MCA. G-banded karyotype analysis should be reserved for patients with obvious chromosomal syndromes (e.g., Down syndrome), a family history of chromosomal rearrangement, or a history of multiple miscarriages.

Very recently, microarray CGH-based screening revolutionized the area of preimplantation genetic screening. One of the presumed factors causing low pregnancy rates in medically assisted reproduction is the presence of numeric chromosome abnormalities in preimplantation embryos. Preimplantation genetic screening is currently proposed for couples with repeated implantation failure following assisted reproduction treatments to enhance pregnancy success with transfer of normal karyotypic embryos; couples with repeated unexplained miscarriages; and women of advanced maternal age, in order to avoid chromosomally abnormal offspring and to improve the success of *in vitro* fertilization procedures. Comprehensive chromosomal analysis of single cells from human preimplantation embryos using whole-genome amplification and single-cell (polar body, blastomeric, blastocyst cells) comparative genomic hybridization was introduced about 10 years ago. Until very recently, the technology was time consuming, and biopsied embryos needed to be frozen pending the results of the CGH, with associated compromise in embryo quality. However, microarray-based methods will yield similar results and may be sufficiently rapid to permit comprehensive screening without the need for embryo cryopreservation. The Cambridge, UK-based company, BlueGenome, has elaborated an array-based CGH protocol which allows for analysis of all chromosomes from biopsied polar bodies, blastomers, or blastocyst cells within 11 to 13 h. The technology was validated in different clinical trials, and was proven to be the best and the fastest method for a chromosomal aneuploidy detection tool in preimplantation genetic screening.

5.2.5   *Expression arrays*

Microarray or DNA chip technology has been used for many applications such as comparative genomic DNA hybridization, genome resequencing with the tiling array, or ChIP-on-chip to identify interactions between proteins and genomic DNA (gDNA). However, the most common application of this technology is gene expression profiling. The basis of a microarray is hybridization between complementary nucleic acids. Although the principle of nucleic acid hybridization is not new, microarrays have opened the way for the parallel detection and analysis of the patterns of expression of thousands of genes in a single experiment [35]. The targets are labeled and the array contains a large number of probes (up to 10^6) that are attached to a solid support.

The most common microarray format is a collection of DNA samples covalently attached to a stable substrate such as a glass slide. Spots of DNA (printed DNA sequences) each represent a specific gene (or part of a gene) and are arranged in an orderly pattern across the solid surface by robotic technologies in gene expression profiling; the targets are comprised of a population of complementary DNA (cDNA) or complementary RNA (cRNA) sequences that are copies of the messenger RNA (mRNA), labeled with fluorochromes. The targets are applied directly to the microarray and hybridized overnight. After the hybridization step, the microarrays are scanned and images are generated. The signal intensity is proportional to the relative abundance of the target present in the starting nucleic acid sample and is quantified using a high-resolution scanner. Statistical and functional analyses are performed on the generated data. However, the microarray is a powerful tool; its analysis is subjected to reproducibility problems and integration of large-scale data. To compare data obtained from different microarrays of the same experimental analysis, it is necessary to normalize data. After data normalization, statistical analyses allow analyzing data with good confidence [36].

Previous studies have aimed to identify sets of individual genes whose differential expression is highly correlated with phenotypic changes (e.g., genes that may be over- or underexpressed in cancer relative to normal). In these cases, increased or decreased absolute mRNA concentration levels above some threshold are put forth as candidates for disease-induced (or causing) perturbations. Unfortunately, the statistically significant genetic changes often depend largely on the context of the microarray experiment. Even when thresholds are tuned to produce statistically significant results, findings can depend heavily on a number of factors, such as the experimental design and the type of data normalization. This often limits the information that can be gleaned from the expression patterns of individual genes. As an alternative approach, studying gene expression in the context of networks may yield greater insight into mechanisms and functional changes associated with disease. Recently, methods for analyzing microarray data have focused not on individual genes, but instead gene sets characterizing biologically meaningful pathways or networks [37].

The number of techniques available to improve microarray experiments has increased rapidly for studies that examine the bacterial side of the host–pathogen interaction. While classic molecular biology experiments focus on selected genes, microarrays provide detailed knowledge of host–pathogen interactions by high-throughput whole-genome analyses. Microarrays allow the analysis of bacterial genomic content and the identification of genes involved in host–pathogen interactions. Microarrays can also be used to highlight cross-gene regulation between bacteria and their host during the course of an infection [36]. Moreover, these techniques have provided insights into how obligate intracellular bacteria adapt to the host cell environment by targeted genomic variations and transcriptional regulation of specific genes [36].

Another important field in microarray studies is the expression profiling of a set of clinical samples, which can identify potential drug targets. Once a particular target is shown to be robustly associated with the disease, it becomes a target for compound screening. Expression profiling can play a role here if, for example, a specific gene expression response can be attributed to the specific target. In subsequent compound validation and optimization stages, expression analysis can help identify the most potent version of a drug. In the area of safety assessment, animal models and/or relevant cell lines (particularly liver cells) are exposed to the drug, and here gene expression profiling can be highly informative. Finally, in both early and late stage clinical trials, expression biomarkers derived from the earlier microarray studies, or even whole-transcriptomic arrays, may be used to monitor drug efficacy and any potential *in vivo* toxicity. Identifying a relevant set of biomarkers will reliably indicate the presence of the disease.

There are general difficulties that can be encountered in microarray-based disease studies. Samples may derive from a variety of sources: direct surgical biopsy, postmortem samples, cell lines, or even samples derived from fixed paraffin sections; biopsy samples may not be homogeneous, and they may contain multiple cell types, genetic heterogeneity between different patient samples. From the above, it should be obvious that issues of experimental design, primary data analysis, and statistical method selection are paramount in disease studies. It can cost millions of dollars to develop a diagnostic test and hundreds of millions to develop a drug; if such developments are founded on poor primary data, the waste of valuable R&D money is obvious.

Taken together, the prospects for developing new effective treatments based on these genomics studies are brighter than with the more limited insights coming from traditional diagnostic tools. Gene expression profiling focusing on the prognostic genes identifies groups of patients that are more likely to respond well to current treatment regimes and also flags a set of patients that are unlikely to benefit from these treatments.

The increasingly mature gene expression platforms now available allow a comprehensive view of the transcriptome in virtually any situation where a cell population can be isolated. With several years of experience in terms of probe design, normalization methods, application of robust statistical metrics, and data exploration tools, the field is beginning to stabilize. Problems of *interplatform* comparability are becoming issues of the past with the ongoing development of appropriate internal and external controls. While it is clear that developments will continue, particularly in the areas of data processing and analysis, it is a little over 10 years since the first cancer gene expression study, and the changes in terms of array platforms and their analysis have been dramatic. It is to be hoped that the next decade will provide equally impressive improvements [38].

5.2.6 Noncoding RNA

The noncoding RNAs (ncRNAs) and their functions were out of focus of investigations, primarily due to the fact that they were not converted into proteins and therefore the interpretation of their function was difficult. This information, which was thought to be encoded in the *junk* sections of the genome, needs to be distinguished from *transcriptional noise*. According to the estimation, the nonprotein-coding RNA can share 97% to 98% of the transcriptional output of a human cell, and the advances in large-scale sequencing confirmed that the majority of transcriptome has no or little protein-coding ability. After the discovery of RNA interference and microRNAs (miRNAs), which met with success in the last decade, the approach to nonprotein-coding genomic sections changed fundamentally and the life science began to puzzle out a new, defined layer of gene expression regulation. Meanwhile investigations of small noncoding RNAs unfolded many, yet unpredictable regulatory relationships. It is now clear that functions of the noncoding kingdom involve gene silencing, gene transcription, DNA imprinting, DNA demethylation, chromatin structure dynamics, and RNA interference, as well. At the same time, several efforts were induced to unveil new super-families of noncoding genes in the genomes with known sequence.

Noncoding RNAs are summarized as transcripts that are not translated (synonym: nonprotein-coding RNA). They can be divided into small (small nucleolar RNAs [snoRNAs], Piwi interacting RNAs [piRNAs], etc.) and medium/large size molecules, and thanks to their popularity, miRNAs are often classified as an independent group (review in Costa [39]). Another classification of RNA families integrates the well-known ribosomal, transfer, and messenger RNAs as housekeeping RNAs in spite of their regulatory class. Beside the dominant presence of small ncRNAs like miRNAs in the literature, even more efforts target the recognition of function of long noncoding RNAs (longer than 200 nt). Although there are gaps in the understanding of long noncoding RNAs, their revealed functional spectrum is surprisingly wide, as they are involved in regulation of gene expression on chromatin (e.g., recruitment of chromatin remodeling complex PCG2 by HOTAIR) transcriptional (e.g., repressive activity of stress-induced expression of ncRNA associated with the cyclinD1-promoter) and posttranscriptional levels (either by being processed into endo-siRNA, or by influencing the pattern of splicing); moreover they can modulate protein activity (e.g., NRON suppresses the nuclear accumulation of NFAT) [40].

Among nonprotein-coding RNAs, the key role of microRNAs has to be emphasized in regulating gene expression by interacting with messenger RNA (mRNA)—either by inhibiting mRNA translation [41,42] or by causing mRNA degradation [43,44]. Mature forms of miRNAs are a class of conserved, approximately 21–25 nucleotide-long RNAs, which are coded in individual transcription units (intergenic, intronic, or exonic). After a characteristic maturation process, miRNAs integrate into the RISC complex and guide it to the 3′ UTR of target messenger RNAs and induce its degradation or inhibit the translation of the protein. The interaction between miRNAs and target mRNA depends on the sequence complementarity. MiRNAs have been identified in both plants and animals, acting as key regulators of multicellular differentiation. In accordance with miRNA's control, a range of developmental events, such as timing of cell-fate decision, stem cell maintenance, apoptosis, and organ morphogenesis, are suggested to play a role in the appropriate establishment of the tissue or cell type-specific expression patterns along the development. It is now established that miRNAs contribute not only to developmental process-related functions, as they are able to adjust the concentration of many proteins in a synchronized and rapid manner, but miRNAs are also thought to play an important role in cell responses and adaptation to the environment. Moreover, miRNAs can function as important tools in the communication of cells. As components of the delivered *exosomal shuttled RNA*, they can be detected in exosomes secreted by mast cells. Through this newly discovered form of cell–cell communication, the donor cell may modulate the posttranscriptional system of target cells directly [45].

As the region of complementarity is typically short and the base pairing imprecise between animal miRNA and their targets, a given miRNA can own hundreds of targets. The result of miRNA-mediated regulation can be detected mainly on the protein level in animals, and for this purpose, application of high-throughput experimental techniques is needed. To resolve this problem, several computational algorithms were developed, which are able to predict the putative targets of miRNAs more or less reliably. These algorithms' search sequences are called *miRNA seeds*, as complementary motifs in the 3' UTR of genes. Basically, these algorithms are working with parameters including sequence complementarity, stability of heteroduplex, and evolutional conservation; however, the even small differences of algorithms can cause a great shift in the list of targets. In fact, the extent of overlap among the target sets predicted by different algorithms is typically low, and among the predicted hundreds of targets, only a small proportion were experimentally validated [46]. Although the prediction of interactions between RNA molecules by using pure sequence data seems to be more feasible compared to transcription factor-DNA predictions, more, as yet unidentified factors should be taken into consideration, since the overall performance is typically just acceptable with approximately two-thirds of the false-positive rate determined by high-throughput proteomic analysis [47,48]. Hence, computational modeling of miRNA–mRNA interactions and accurate prediction of miRNA target sites (with reduction of the number of false positives) remains an important and open question to bioinformatics.

The experimental observations [49,50] and bioinformatic predictions begin to delineate some basic principles in miRNA-driven regulation. Interestingly, the miRNA target sites in the genes of ubiquitous, basic metabolic processes are relatively underrepresented, while the genes involved in cell growth/death, development (especially, e.g., in the axon guidance pathway), transcriptional regulation, and intercellular communication are exposed to miRNA-driven control more frequently. Possibly, the miRNA-mediated regulation might be necessary in processes when, besides the fundamental regulation, a fine balancing is also required for driving the cell to the most proper response for different signals [51].

The binding sites of a given miRNA are present in several, functionally related transcripts, and this coregulatory principle may allow for miRNAs to harmonize the elements of a pathway or related pathways. There is some evidence of the coordinative action, while different miRNAs own independent binding sites in the 3' UTR of the same mRNA. In this case, miRNAs are able to influence the level of their targets in a coordinative manner, increasing the degree of translational repression synergistically (Figure 5.2). In accordance with this observation, the number of binding sites correlates with the effectiveness of translational repression [49].

Furthermore, the overlapping binding sequences in the 3' UTR can result in a competition between the respective miRNAs [50]. Finally, more separate binding sites do not mean that all of the silencers are expressed and present at the same time. Under certain circumstances and in different cell types, a given combination of miRNAs can be observed, which provides various choices in setting up the proper level of target. This

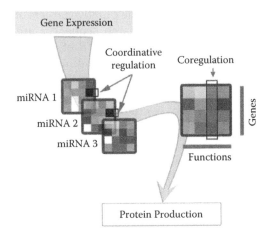

Figure 5.2 Integration of microRNA target pattern as a new layer of regulation of gene expression.

latter principle is supposed to be called *pattern differential regulation* [50]. The various recognized phenomena can be the basis of interpretation of the experimental results concerning miRNAs and the integration as a new layer to the knowledge of regulation of gene expression.

The continuous development of experimental tools (like next-generation sequencing, transgenic animals, and high-throughput proteomics) with the contribution of bioinformatics will allow looking into the regulatory networks complemented by ncRNA. Although, among the expanding number of ncRNA genes, the class of miRNAs can supply the most eminent examples in this emerging research field presently, the discovery of new representatives and classes of noncoding RNAs can surely be anticipated.

5.2.7   A current application: Immunomics

Immunomics offers an excellent example for application of the systems biology and modern genomics/proteomics technology. Immunomics comprises a fruitful combination of modern molecular immunology, animal modeling, omics technologies, and application of ever-increasing immune databases.

Immunomics, the combination of modern immunology and informatics, has the power of an integrated approach to characterize master cytokines, their receptors, signaling pathways, and identification of novel components of innate immunity [1].

Clinical immunomics covers a broad range of diseases and this volume is an attempt to bring together work representing a spectrum of immunomic applications to present key technologies that will shape clinical immunology in the near future [52]. Examples are coming that underline the relevance of immunomics to cancer, autoimmunity, allergy, and primary immunodeficiencies, along with those that address more basic aspects of large-scale studies of immunology. Various combinations of bioinformatics, biostatistics, instrumentation, molecular biology, animal models, and clinical samples serve the development of clinical immunology. The combination of these approaches enables identification of complex association networks between immunologic and clinical outcomes.

Databases (see more in Section 4.1 in Chapter 4), for example, the concept of the Essential Human Immunome and informatics, are valuable resources for storage and computational analysis of genes and proteins of the immunome [53]. Excellent examples are provided by the immunodeficiency data banks focusing on genetic variations found in primary immunodeficiencies and the analysis of related immunome entries from some 5000 patients. These resources assist health professionals to select suitable genetic and clinical tests for immunodeficiencies.

Allergy immunomics combines allergology, structural biology, and bioinformatics. The Structural Database of Allergenic Proteins (SDAP) [54] database offers a detailed review of practically all known plant allergens, their structure, allergenicity, and allergic cross-reactivity. Grouping of allergens into structural families enables identification of shared molecular properties of allergens and the basis for the prediction of allergenicity and analysis of cross-reactivity.

The likely major advances of immunomics will be seen first in the fields of developing vaccines and high-throughput diagnostics. In vaccine science, the combination of genetics (at least at three levels with genetic heterogeneity-proteasomal processing, peptide transport to ER, and MHC-binding [Figure 5.3] bioinformatics prediction tools) and a set of experimental models (biochemical assays and animal models) together give rise to establishing a new personalized vaccination strategy [55]. There are at least three target points with genetic diversity (MHC, proteasome, TAP) for individual considerations when establishing *intelligent* vaccines [6,11].

Three case studies, including influenza, tularemia, and human papilloma virus, are presented to demonstrate the utility of immunomics for epitope-based subunit vaccine development [56]. Spontaneously developed mammary tumors enabled the optimization of vaccine scheduling and minimization of the number of vaccine injections [56]. In autoimmune diseases, immunomics helps in emerging therapies for treatment of myasthenia gravis. New diagnostic and therapeutic options for multiple sclerosis, a multifactorial inflammatory autoimmune disease, is promised through combined immunomics strategy [3,4]. The immunomics

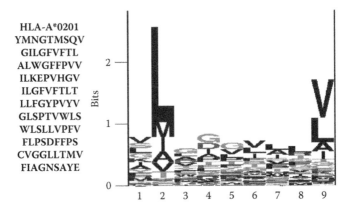

Figure 5.3 Binding prediction of the amino acid sequence of a nonameric peptide to a genetically determined MHC grove (e.g., HLA-A*0201).

approach combines multiple approaches: molecular biology for blocking tissue damage through blocking the activity of matrix metalloprotein-ases, and nanotechnology for formulation and delivery of therapeutic agents, combined with healthy and functional foods. The holistic nature of immunomics is where the latest technological advances in combination with lifestyle (diet) modification can exert a profound effect on the course of an autoimmune disease.

In cancer immunology, glycomics as a tool was applied [5]. Early detection and therapeutic responses use gangliosides, a class of glycoanti-gens which are overexpressed on the tumor cells, relative to their expres-sion on normal cells. Gangliosides are released from tumor cells into both the tumor microenvironment and the bloodstream. They act as immu-nomodulatory agents in the tumor microenvironment while in the blood they can be used as biomarkers for diagnostics and prognostics in oncol-ogy. Because they are recognized as immunogens, gangliosides represent suitable targets for both diagnosis and therapy.

The concept of translational immunomics and its applications in immunoprevention of cancer is emerging; a combination of genomics and mouse models of mammary tumor for efficient identification of onco-antigens, tumor profiling, and immune monitoring. The combination of mathematical modeling and *in vivo* immunization with the triplex (tri-component) vaccine of mice may result in a principal breakthrough in combined tumor therapies.

Identification of clinical outcome features and their precise measure-ment is essential for decision making on selection of optimal therapy and modification of therapy by assessment of early responses [10]. Translation of technological and scientific advances into actual clinical solutions remains a huge task and immunomics offers the means for systematic analysis and speeding up the selection, discovery, design, and validation of new diagnostic and therapeutic products for the 21st century.

5.3 Computational biology/bioinformatics

5.3.1 Database searching

The exponential growth of experimental results promises a more and more thorough understanding of biological processes. Given the enormous amount of data, it is challenging to parse chemical and biological information efficiently. The common method of handling large sets of data is storing them in databases, and retrieving only the data of our interest. Storage of data in relational databases (in which data are structured by categories) is a widely used and easy-to-manage way of representing biological data sets [57]. Databases, however, also represent a new approach toward information management. Searching for literature in biological databases or simply trying to find the protein that belongs to a specific sequence with the Basic Local Alignment Search Tool (BLAST), are all requests for the closest matching entries of a database.

Corresponding to the diversity of experimental settings, there are many possible ways to categorize biological data. When searching for a particular answer, it is important to know which of the databases can help us. Comprehensive databases about all existing databases, like DBD [58] and MetaBase [59], or journals focusing on databases, like the yearly database issue of *Nucleic Acid Research* [60], may prove to be useful. It is practical to group databases by their main field of interest. However, it is almost impossible to enumerate all types of databases because of the expanding cohort of individual scientific subspecialties that take advantage of the structured knowledge. Some of the most important types include chemical structure databases like ChemSpider [61], nucleic acid databases like GenBank [62], genomic databases like Ensembl [63], protein information databases like SwissProt/UniProt [64], protein family databases like Pfam [65], protein structure databases like PDB [66], phylogenic databases like TreeBASE [67], immunological databases like IMGT [68], carbohydrate databases like GlycoSuiteDB [69], metabolic pathway databases like MetaCyc [70], gene expression databases like GEO [71], taxonomic databases like ITIS [72], and bibliographic databases like PubMed [73].

Another important aspect is how processed the information is in the database. Databases are often categorized as primary and secondary ones [74]. Primary databases contain only raw experimental results of a certain kind. As high-throughput technologies evolve in all fields of life sciences, the number of primary databases increases day to day. Secondary databases contain the analysis of raw data found in primary databases. Even if it is possible to calculate some of these parameters by analyzing the data of primary databases, it is much less time-consuming to store the results in a dedicated depository for immediate retrieval.

When searching for information in databases, however, one usually wants to retrieve the most information about a single entry. Because of storing biological data in specialty-specific databases, the same query may be executed many times for different databases. The competing repositories of the same or similar topic also increase the redundancy [75]. The latter issue is solved by cooperation among databases, like DDBJ [76] (Japan), GenBank [77] (United States), and EMBL [78] Nucleotide Sequence Database (Europe) exchanging new and updated data on a daily basis [79].

Meta-databases however, offer a more powerful means of accessing the most relevant and most comprehensive data available about an entity [80]. Being more than just another database of databases, after performing the subsequent searches in each separate database, the results are integrated into one output. One of the most frequently used meta-databases is Entrez [81], providing a full spectrum of information from raw experimental data to the relevant literature. Other meta-databases like Diana [82], IEDB [83], Influenza Virus Resource [84], and Magia [85], to mention some widely used resources in different areas like miRNA research, phylogeny, and immunology, combine information extracted from experimental data sets with prediction algorithms to offer complex analysis of interactions of the queried biological process [86]. Pathway analysis with the help of databases may shed light on previously unexplored interactions and molecular processes. This promising opportunity makes it easy to understand why online bioinformatics tools, combining prediction and database searching, become more and more popular [87].

As most databases are Web based to provide convenient access and up-to-date information, an easy-to-use interface is also available. Although it is user friendly, when performing high-scale bioinformatics studies, it is necessary to automate a search and integrate it into the analysis pipeline. This automated approach offers the opportunity of meeting more complex demands. Without having to set out for complicated data mining strategies, simple queries can be automated in script-friendly databases. Currently, the two most common script languages are Python and Perl. Since the availability of open-source code collections BioPython and BioPerl, both languages make it possible to combine easy data retrieval from the most important biological and chemical databases with high-level programming.

The constantly growing scientific literature may be considered a regularly updated (but semistructured) database. By extracting and structuring current scientific knowledge using data-mining algorithms, it is possible to answer complex questions. Open access databases, on the other hand, represent a novel (and strictly structured) means of donating scientific information to the public—thus expanding the information content of scientific literature.

5.3.2 Bayesian methods from study design through data analysis to decision support

After the revolutionary development of computing hardware, computer science, and computer intensive simulation in the second half of the 20th century, the hallmark of the new century/millennium is the flood of electronically accessible data. Thanks to the unexpected development of measurement techniques, data pervades and transforms traditionally data-poor, knowledge-rich sciences such as biology and medicine. However, compared to astrophysics or business, the status and role of data in biomedicine is unique in many respects, such as the presence of multiple, weakly connected levels and the possibility of *completeness* of an observation of a given level, that is, the –omics.

The enormous dimensionality of each level, the heterogeneity of the levels, and the richness of the *a priori* knowledge create an entangled threefold challenge throughout the cycle of scientific discovery: in the study design, in the data analysis, in the interpretation/translation, and in the application. One may witness important developments in each of these phases, namely (1) adaptive study design, (2) biomarker analysis and inference of model properties in systems biology, (3) approaches to fusion of data and knowledge for translational research, and (4) decision support.

Adaptive or flexible or sequential statistical study designs have a long history in clinical trials, particularly in pharmacology [88,89]. Because of the financial, medical, and other costs of collecting samples, adaptive study design traditionally focused on the adaptive selection of sample size, assuming a fixed set of measured or observed variables, and fixed statistical tests. The Bayesian framework offers a principled approach to sequential study design, which fits the knowledge-rich biomedical context of biomedicine by allowing informative priors and informative utilities [90]. This biostatistical approach involves multivariate extensions within the framework of active learning for optimizing the observed variables and samples as well [89,90].

In the omics era, a particularly challenging task is the biomarker selection or the feature subset selection problem, that is, to find the associated/predictor variables for single or multiple outcomes. The refinement of the concept of association, such as weak, strong, conditional, and direct, along with its causal extensions in systems biology, such as probabilistic, direct, and causal effect modifiers is needed. The semantics of Bayesian networks, which provides a formal language and graphical visualization of these concepts is a central issue. This model class was extended to analyze structured data, such as hierarchical or temporal data. One also has to consider the robustness of the Bayesian statistical framework and model averaging against multiple testing and model complexity.

To illustrate approaches to solving the interpretational/translational bottleneck, the advantages of the joint analysis of heterogeneous data sets and the set-based analysis are discussed. Various preprocessing and post-processing aggregations, such as *horizontal aggregation*, which aggregates variables along a specific effect (e.g., through effected pathways), and *vertical/heterogeneous aggregation*, which aggregates heterogeneous effects of a variable from multiple omics levels (i.e., analyzing jointly heterogeneous omics data sets) are other candidates to cope with the high number of variables. Moreover, the possibility of *post hoc* meta-analysis based on probably approximately correct characterizations of the data sets is another option to support the interpretation.

Finally, the usage of a probabilistic causal model for sequential diagnosis and treatment selection has to be evaluated, illustrating concepts such as value of information and cost–benefit analysis.

References

1. Béné, M. C., Stockinger, H., Capel, P. J., Chapel, H., and Knapp, W. 2000. Clinical immunology: A unified vision for Europe. *Immunol. Today* 21:210–211.
2. Falus A. (ed). 2005. *Immunogenomics and Human Disease*. New York: Wiley.
3. Purcell, A. W. and Gorman, J. J. 2004. Immunoproteomics: Mass spectrometry-based methods to study the targets of the immune response. *Mol. Cell. Proteomics* 3:193–208.
4. Brusic, V., Marina, O., Wu, C. J., and Reinherz, E. L. 2007. Proteome informatics for cancer research: From molecules to clinic. *Proteomics* 7:976–991.
5. Schönbach, C., Ranganathan, S., and Brusic, V. (eds.). 2007. *Immunoinformatics*. New York, Heidelberg: Springer.
6. Tegnér, J., Nilsson, R., Bajic, V. B., Björkegren, J., and Ravasi, T. 2006. Systems biology of innate immunity. *Cell Immunol.* 244:105–109.
7. Li Pira, G., Kern, F., Gratama, J., Roederer, M., and Manca, F. 2007. Measurement of antigen specific immune responses: 2006 update. *Cytometry B Clin. Cytom.* 72:77–85.
8. Sachdeva, N. and Asthana, D. 2007. Cytokine quantitation: Technologies and applications. *Front. Biosci.* 12:4682–4695.
9. Harnett, M. M. 2007. Laser scanning cytometry: Understanding the immune system *in situ*. *Nat. Rev. Immunol.* 7:897–904.
10. Tremoulet, A. H. and Albani, S. 2005. Immunomics in clinical development: Bridging the gap. *Expert Rev. Clin. Immunol.* 1:3–6.
11. Brusic, V. and August, J. T. 2004. The changing field of vaccine development in the genomics era. *Pharmacogenomics* 5:597–600.
12. Zare, H., Shooshtari, P., Gupta, A., and Brinkman, R. R. 2010. Data reduction for spectral clustering to analyze high-throughput flow cytometry data. *BMC Bioinformatics* 11:403–416.

13. James, P. 1997. Protein identification in the post-genome era: The rapid rise of proteomics. *Q. Rev. Biophys.* 30: 279–331.
14. Wilkins, M. R., Pasquali, C., Appel, R. D., Ou, K., Golaz, O., Sanchez, J-C., Yan, J. X., Gooley, A. A., Hughes, G., Humphery-Smith, I., Williams, K. L., and Hochstrasser, D. F. 1996. From proteins to proteomes: Large-scale protein identification by two-dimensional electrophoresis and amino acid analysis. *Nature Biotech.* 14:61–65.
15. Perfetto, S. P., Chattopadhyay, P. K., and Roederer, M. 2004. Seventeen-colour flow cytometry: Unraveling the immune system. *Nature Rev. Immunol.* 4:648–655.
16. Mahnke, Y. D. and Roederer, M. 2007. Optimizing a multi-colour immuno-phenotyping assay. *Clin. Lab. Med.* 27:469–485.
17. De Rosa, S. C., Herzenberg, L. A., and Roederer, M. 2001. 11-color, 13-parameter flow cytometry: Identification of human naive T cells by phenotype, function, and T-cell receptor diversity. *Nat. Med.* 7:241–248.
18. Ibrahim, S. F. and Engh, G. 2003. High-speed cell sorting: Fundamentals and recent advances. *Curr. Opin. Biotech.* 14:5–12.
19. Morgan, R., Varro, H., Sepulveda, J. A., Ember, J., et al. 2004. Cytometric bead array: A multiplexed assay platform with applications in various areas of biology. *Clin. Immunol.* 110:252–266.
20. Kellar, K. L. and Douglass, J. P. 2003. Multiplexed microsphere-based flow cytometric immunoassays for human cytokines. *J. Immunol. Methods* 279:277–285.
21. Iannone, M. A., Taylor, J. D., Chen, J., Li, M. S., Rivers, P., Slentz-Kesler, K. A., and Weiner, M. P. 2000. Multiplexed single nucleotid polymorphism genotyping by oligonucleotide ligation and flow cytometry. *Cytometry* 39:131–140.
22. Clutter, M. R., Krutzik, P. O., and Nolan, G. P. 2005. Phospho-specific flow cytometry in drug discovery. *Drug Discov. Today Technol.* 2:295–302.
23. Krutzik, P. O., Hale, M. B., and Nolan, G. L. 2005. Characterization of the murine immunological signaling network with phospho-specific flow cytometry. *J. Immunol.* 175:2366–2373.
24. Pennisi, E. 2007. Breakthrough of the year: Human genetic variation. *Science* 318:1842–43.
25. National Human Genome Research Institute: http://www.genome.gov/26525384.
26. Barrett, J. C. et al. 2009. Genome-wide association study and meta-analysis find that over 40 loci affect risk of type 1 diabetes. *Nat. Genet.* 41:703–707.
27. Wellcome Trust Case Control Consortium: http://www.wtccc.org.uk/ccc2/.
28. McClellan, J. and King, M. C. 2010. Genetic heterogeneity in human disease. *Cell* 141:210–217.
29. Antal, P., Millinghoffer, A., Hullám, G., Hajós, G., Szalai, C., and Falus, A. 2009. Bioinformatic Platform for a Bayesian, Multiphased, Multilevel Analysis in Immunogenomics. In *Bioinformatics for Immunomics*, ed. D. R. Flower, M. N. Davies, and S. Ranganathan, 157–185. New York, Heidelberg: Springer.
30. Oak Ridge National Laboratory: http://www.ornl.gov/sci/techresources/Human_Genome/home.shtml 2009.
31. Rusk, N. and Kiermer, V. 2008. Primer: Sequencing—The next generation. *Nat. Methods* 5:15.

32. Venter, J. C. 2010. Multiple personal genomes await. *Nature* 464:676–7.

33. Database of Genomic Variants: http://projects.tcag.ca/variation.

34. https://isca.genetics.emory.edu.

35. Bryant, P. A., Venter, D., Robins-Browne, R., and Curtis, N. 2004. Chips with everything: DNA microarrays in infectious diseases. *Lancet Infect. Dis.* 4:100–111.

36. Leroy, Q. and Raoult, D. 2010. Review of microarray studies for host-intracellular pathogen interactions. *J. Microbiol. Methods* 81(2):81–95.

37. Eddy, J. A., Sung, J., Geman, D., and Price, N. D. 2010. Relative expression analysis for molecular cancer diagnosis and prognosis. *Technol. Cancer. Res. Treat.* 9(2):149–159.

38. Russell, S., Meadows, L. A., and Russell, R. R. 2009. *Microarray Technology in Practice*. Oxford: Elsevier.

39. Costa, F. F. 2007. Noncoding RNAs: Lost in translation? *Gene* 386:1–10.

40. Mercer, T. R., Dinger, M. E., and Mattick, J. S. 2009. Long noncoding RNAs: Insights into functions. *Nat. Rev. Genet.* 10:155–159.

41. Ambros, V. 2004. The functions of animal microRNAs. *Nature* 431:350–355.

42. Bartel, D. P. 2004. MicroRNAs: Genomics, biogenesis, mechanism, and function. *Cell* 116(2):281–297.

43. Krutzfeldt, J., Rajewsky, N., Braich, R., Rajeev, K. G., Tuschl, T., Manoharan, M., and Stoffel, M. 2005. Silencing of microRNAs *in vivo* with "antagomirs." *Nature* 438:685–689.

44. Lim, L. P., Lau, N. C., Garrett-Engele, P., Grimson, A., Schelter, J. M., Castle, J., Bartel, D. P., Linsley, P. S., and Johnson, J. M. 2005. Microarray analysis shows that some microRNAs downregulate large numbers of target mRNAs. *Nature* 433:769–773.

45. Valadi, H., Ekstrom, K., Bossios, A., Sjostrand, M., Lee, J. J., and Lotvall, J. O. 2007. Exosome-mediated transfer of mRNAs and microRNAs is a novel mechanism of genetic exchange between cells. *Nat. Cell Biol.* 9:654–659.

46. Rajewsky, N. 2006. MicroRNA target predictions in animals. *Nat. Genet.* 38:S8–13.

47. Baek, D., Villen, J., Shin, C., Camargo, F. D., Gygi, S. P., and Bartel, D. P. 2008. The impact of microRNAs on protein output. *Nature* 455:64–71.

48. Selbach, M., Schwanhausser, B., Thierfelder, N., Fang, Z., Khanin, R., and Rajewsky, N. 2008. Widespread changes in protein synthesis induced by microRNAs. *Nature* 455:58–63.

49. Doench, J. G., Petersen, C. P., and Sharp, P. A. 2003. siRNAs can function as miRNAs. *Genes Dev.* 17:438–442.

50. Hua, Z., Lv, Q., Ye, W., Wong, C. K., Cai, G., Gu, D., Ji, Y., Zhao, C., Wang, J., Yang, B. B., and Zhang, Y. 2006. MiRNA-directed regulation of VEGF and other angiogenic factors under hypoxia. *PLoS ONE* 1:e116.

51. Gaidatzis, D., Van Nimwegen, E., Hausser, J., and Zavolan, M. 2007. Inference of miRNA targets using evolutionary conservation and pathway analysis. *BMC Bioinformatics* 8:69–79.

52. Falus, A. (ed.). 2009. *Clinical Applications of Immunomics*. New York, Heidelberg: Springer.

53. Ortutay, C. and Vihinen, M. 2006. Immunome: A reference set of genes and proteins for systems biology of the human immune system. *Cell. Immunol.* 244:87–89.

54. SDAP: Structural Database of Allergenic Proteins: http://fermi.utmb.edu/SDAP/.
55. de Groot, A. S., Moise, L., McMurry, J. A., and Martin, W. 2009. Epitope-Based Immunome-Derived Vaccines: A Strategy for Improved Design and Safety. In *Clinical Applications of Immunomics* (ed.) A. Falus, 39–70. New York, Heidelberg: Springer.
56. Pappalardo, F., Motta, S., Lollini, P. L., and Mastriani, E. 2006. Analysis of vaccines' schedules using models. *Cell. Immunol.* 244:137–40.
57. Wesley, D. 2000. Relational database design. *J. Insur. Med.* 32:63–70.
58. DBD: Database of Biological Database: http://www.biodbs.info.
59. http://biodatabase.org/index.php.
60. Nucleic Acids Research: http://www3.oup.co.uk/nar/database/c.
61. http://www.chemspider.com
62. NCBI: Nucleotide: http://www.ncbi.nlm.nih.gov/sites/entrez?db=nucleotide.
63. Ensembl: http://www.ensembl.org/index.html.
64. UnitProt: http://www.uniprot.org.
65. Sanger Institute: http://pfam.sanger.ac.uk.
66. Protein Data Bank: http://www.rcsb.org/pdb/home/home.do.
67. TreeBASE: http://www.treebase.org/treebase-web/home.html.
68. http://imgt.cines.fr.
69. GlycosuiteDB: http://glycosuitedb.expasy.org/glycosuite/glycodb.
70. MetaCyc: http://metacyc.org.
71. NCBI: Gene Expression Omnibus: http://www.ncbi.nlm.nih.gov/geo.
72. Integrated Taxonomic Information System: http://www.itis.gov.
73. PubMed: http://www.ncbi.nlm.nih.gov/pubmed.
74. Altman, R. B. 2004. Building successful biological databases. *Brief. Bioinformatics* 5:4–5.
75. Burgun, A. and Bodenreider, O. 2008. Accessing and integrating data and knowledge for biomedical research. *Yearb. Med. Inform.* 91–101.
76. DNA Data Bank of Japan: http://www.ddbj.nig.ac.jp/searches-e.html.
77. NCBI: Nucleotide: http://www.ncbi.nlm.nih.gov/sites/entrez?db=nucleotide.
78. UniProt: www.ebi.ac.uk/uniprot.
79. Hermjakob, H. and Apweiler, R. 2002. TEMBLOR—Perspectives of EBI database services. *Comp. Funct. Genomics* 3:47–50.
80. Sullivan, D. E., Gabbard, J. L., Jr., Shukla, M., and Sobral, B. 2010. Data integration for dynamic and sustainable systems biology resources: Challenges and lessons learned. *Chem. Biodivers.* 7:1124–41.
81. NCBI: Entrez, The Life Sciences Search Engine: http://www.ncbi.nlm.nih.gov/Entrez.
82. DIANA Lab: http://diana.cslab.ece.ntua.gr.
83. Immune Epitope Database and Analysis Resource: http://www.immuneepitope.org.
84. NCBI: Influenza Virus Resource: http://www.ncbi.nlm.nih.gov/genomes/FLU/FLU.html.
85. Magia: http://gencomp.bio.unipd.it/magia/start.
86. Kandpal, R., Saviola, B., and Felton, J. 2009. The era of "omics unlimited." *Biotechniques* 46:351–2.
87. Friedman, F. 2004. Inferring Cellular Networks Using Probabilistic Graphical Models. *Science* 303:799–805.

88. Pearl, J. 2000. *Causality: Models, Reasoning, and Inference.* Cambridge: Cambridge University Press.
89. Stephens, M. and Balding, D. J. 2009. Bayesian statistical methods for genetic association studies. *Nature Rev. Genetics* 10:681–690.
90. Antal, P., Millinghoffer, A., Hullám, G., Szalai, C., and Falus, A. 2008. A Bayesian View of Challenges in Feature Selection: Feature Aggregation, Multiple Targets, Redundancy and Interaction, ECML/PKDD (Workshop on new challenges for feature selection in data mining and knowledge discovery, 2008) (FSDM08), JMLR: Workshop and Conference Proceedings, Bioinformatics Workgroup, BUTE-DMIS. 4:74–89.

chapter six

Novel chemogenomic approaches to drug design

Didier Rognan

Contents

6.1   What is chemogenomics?

For a long time, drug designers focused on a single macromolecular target and a single or very few chemical series (Wermuth, 2006). The selectivity of preclinical candidates for the intended target was only addressed at a late stage by profiling the compound against neighboring targets (e.g., receptor subtypes). Therefore, a significant attrition rate in clinical trials (Schuster, Laggner, and Langer, 2005) has been due in the last decades to the unexpected binding of drug candidates to additional targets, either off-targets (Shoshan and Linder, 2008) or antitargets (Klabunde and Evers, 2005) resulting in dubious pharmacological activities, side effects, and sometimes adverse drug reactions (Yang, Chen, and He, 2009). Remarkable advances in structural genomics (Dessailly et al., 2009; Nair et al., 2009) and diversity-oriented chemistry (Nielsen and Schreiber, 2008; Daniel et al., 2009) have changed these practices. On the biological side, the Protein Data Bank (Berman et al., 2000), which stores publicly available three-dimensional (3D) structures of macromolecules, currently stores over 65,000 entries. Outstanding efforts of structural genomic consortia to complete the structural proteome let us anticipate an acceptable coverage of the UniProt database (Schneider et al., 2009) in only 15 years (Nair et al., 2009). On the chemical side, about 27 million unique structures and 435,000 bioactivity screens are available in the PubChem repository (Wang et al., 2009). Mapping pharmacological space in 2006 (Paolini et al., 2006) resulted in more than 1300 targets with significant affinities (< 10 µm) for small molecular-weight ligands. As a consequence, the number of biological data points relating protein to ligand information has exploded in a combinatorial manner. Chemogenomics (Caron et al., 2001) is the discipline at the crossroad of chemo- and bioinformatics, chemistry, biology, pharmacology, and toxicology, which permits analyzing global protein–ligand interaction data and predicting novel links (association, binding) between ligand and target spaces.

Global chemogenomic approaches targeting arrays of ligands (rows) and proteins (columns) to generate huge two-dimensional binding matrices (Figure 6.1) enlarge our vision of how chemical and biological spaces match (Ong et al., 2009).

Experimental chemogenomics is, however, expensive, time consuming, and addresses only a restricted subset of chemical (a few thousand ligands) and biological space (a few hundred targets). Combining bio- and chemoinformatic structural approaches (Paolini et al., 2006; Rognan, 2007; Bajorath, 2008) to fill chemogenomic matrices presents the noticeable advantage of considerably extending space coverage and limiting the number of supporting experimental validations. Predicting missing data in chemogenomic matrices can be done in three ways: (i) column-by-column (search path 1, Figure 6.1) resulting in the virtual screening of

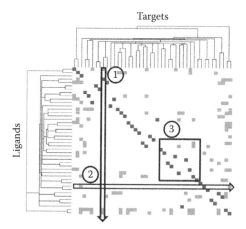

Figure 6.1 The chemogenomic matrix and the three paths (1, 2, 3) to browse it. Targets are classified by rows and ligands by columns. Data here are inhibition constants and colored in green (under a defined threshold) or red (above a defined threshold).

ligand libraries (Shoichet, 2004), (ii) row-by-row (search path 2, Figure 6.1) resulting in the virtual profiling of a ligand against an array of targets (Rognan, 2010), and (iii) in any cross section (search path 3) resulting in true protein–ligand chemogenomic screening (Vert and Jacob, 2008). The present chapter will review recent developments in the field of computational chemogenomics as a follow-up of our previous survey in 2007 (Rognan, 2007).

6.2   How to archive chemogenomic data?

Data archival is a key issue in chemogenomics for the simple reason that experimental data may arise from a wide array of sources and that standardizing chemical and target information (as simple as the name of corresponding molecules) is absolutely necessary to enable clear and nonredundant data storage. For example, there are nine synonyms for aspirin in the PubChem repository (Wang et al., 2009), and five synonyms for the serotonin 5-HT1a receptor in the IUPHAR database (IUPHAR-DB) (Harmar et al., 2009). Efforts to standardize small molecular-weight compounds have led to the International Chemical Identifier (InChI) representation, which is unique to a chemical structure. Analogous attempts to standardize target names have been described, but only for the major therapeutic classes (G-protein coupled receptors, ion channel, nuclear hormone receptors, protein kinases), however, with less success since no gold standard for a target identifier has emerged up to now. The IUPHAR-DB (Harmar et al., 2009) provides a rigorous nomenclature but only for a few target classes. Conversely, the Enzyme Commission attributes a number (EC number) to every enzyme depending on the chemical reaction that it catalyzes. Last, the Universal Protein Knowledgebase (UniProtKB) (Apweiler et al., 2004) gives a unique name to every protein of known sequence. Its Swiss-Prot section is manually annotated and reviewed whereas its TrEMBL section is automatically annotated and not reviewed, thus allowing possible redundancy. Given the difficulty to standardize protein names, the current trend is to label the protein according to the corresponding gene symbol (Scheiber et al., 2009a).

6.2.1 Target-annotated ligand databases

Many biologically annotated ligand repositories are now available from either private or public sources (Table 6.1). These archives differ in the degree of biological annotation (target name, *in vitro* and/or *in vivo* data, ADME/Tox data) and the source of experimental data (scientific journals, patents, high-throughput screening data). Most of these repositories are not public and have been generated by a lengthy and careful reading of literature (medicinal chemistry journals, patents). Depending on the application, qualitative or more quantitative data are necessary. For example, simple QSAR categorization models (binary prediction of protein–ligand complex formation) just need a link between a target and its ligands. Prediction of binding affinities would necessitate storing inhibition constants. Data from cellular and phenotypic screens are not very useful in this context as far as a precise identification of the corresponding target is not available. The number of binding data points that can be exploited is therefore currently not known, notably because a significant redundancy in chemical structures and associated biological data exist among these archives.

Due to financial constraints, many of the commercial databases are not accessible to the academia. Recent initiatives to put high-throughput screening data on the Web are therefore particularly acknowledged from the academic community. Noteworthy are the PubChem (Wang et al., 2009) and ChEMBL (Overington, 2009) databases. PubChem is an open public repository part of the U.S. National Institute of Health (NIH) Molecular Library Roadmap Initiative. It stores biological results from the Molecular Library Screening Center Network but also from other organizations. It currently stores over 28 million compounds, 463,000 bioassays, and is the major source for generating predictive polypharmacology models (Chen, Wild, and Guha, 2009). Although PubChem provides users with a graphical interface, its content is not straightforward to browse and download. Specific stand-alone tools for searching and retrieving PubChem data, notably bioactivity data have flourished in the literature (Hur and Wild, 2008; Chen et al., 2010a). One of the main problems of using PubChem

Table 6.1 Ligand Databases with Biological Annotations

Database	Content	URL	Public
AurSCOPE	Knowledge databases on important target families	http://www.aureus-pharma.com	No
Binding DB	Binding data for 240,000 ligands and 3000 targets	http://www.bindingdb.org	Yes
Bioprint	Profile of 2500 compounds on 160 biological assays	http://www.cerep.com	No
ChemBank	1,266,759 compounds covering 2900 high-throughput screens	http://chembank.broadinstitute.org	Yes
ChemBioBase	Target-centric ligand databases (>1.4 million compounds, 1500 targets)	http://www.jubilantbiosys.com	No
ChEMBL	3 million activity data of 600,000 compounds on 8000 targets	http://www.ebi.ac.uk/chembl	Yes
DrugBank	4800 drug entries and 4500 target information	http://www.drugbank.ca	Yes
GOSTAR	11 million datapoints for 4 million inhibitors	http://www.gostardb.com	No
KKB, ARK	185 data points for 53,000 kinase inhibitors and >200 kinases	http://www.eidogen-sertanty.com	No
IUPHAR-DB	Binding information for 2700 compounds on 600 receptors	http://www.iuphar-db.org	Yes
MDDR	>150,000 biologically active compounds with activity fields	http://www.symyx.com	No
PDSP	48,000 inhibition constants for 1500 compounds	http://pdsp.med.unc.edu/pdsp.php	Yes
PubChem	>28 million structures and 463,000 bioassays	http://pubchem.ncbi.nlm.nih.gov	Yes
Wombat	760,605 biological activities from 309,847 compounds and 1966 unique targets	http://www.sunsetmolecular.com	No

bioassay data is that it is not curated and therefore contains in some instances a large percentage of false positives (Schierz, 2009). In addition, high-throughput screening (HTS) data are usually very imbalanced with a very large proportion of inactives (Li, Wang, and Bryant, 2009), thus enabling a potential difficulty to correctly mine and analyze it. The

ChEMBL database (Overington, 2009) is another interesting archive storing over 3 million activity data on 600,000 compounds and 4360 protein targets. Originally developed by Inpharmatica under the name StARlite, the database was made available to the public in January 2010 and is a premier source for manually curated high quality data (half of data are inhibition constants). A recent quantitative survey of commercial and public ligand databases indicate both a significant degree of overlap but also unique content across all sources, attributable to different strategies for data selection (Southan, Varkonyi, and Muresan, 2009). Out of 2 million compounds extracted from commercial sources, about half of them overlapped with PubChem. Last, the BindingDB database (Liu et al., 2007) is a public, Web-accessible database of measured binding affinities, focusing chiefly on the interactions of proteins considered to be drug targets with small, drug-like molecules. It stores over 540,000 binding data for 240,000 ligands and 3000 targets. Whatever the data set, the user should handle these experimental data with care, notably for deriving QSAR models for several reasons: (i) data from old publications are no more in agreement with current knowledge, notably with respect to receptor subtype selectivity assignments (Figure 6.2a); (ii) most of these archives are biased toward a few targets (e.g., monoaminergic G-protein coupled receptors [GPCRs], protein kinases) which received considerable attention from the pharmaceutical industries the last 20 years (Figure 6.2b); and (iii) functional effects (e.g., agonist, antagonist) are often not documented and may lead to mixing compounds in QSAR models with quite different molecular mechanisms of action.

6.2.2   Ligand-annotated target databases

By ligand-annotated target databases, we mean repositories where the main fields contain pharmaceutically relevant target information (name, sequence, structure) and for which the ligands could be easily retrieved. Such archives are by far less numerous than compound databases for the simple reason that target druggability (Campbell et al., 2010; Schmidtke and Barril, 2010) is more difficult to establish/predict than ligand druglikeness (Ursu and Oprea, 2010).

Three databases (DrugBank, Therapeutic Target Database, SuperTarget) focus on targets addressed by current drugs or compounds under clinical trials and provide overlapping information on the target, their drugs, and therapeutical indications. DrugBank (Wishart et al., 2008) is the premier source for exhaustive information. The database contains nearly 4800 drug entries including >1350 FDA approved small molecule drugs, 123 FDA approved biotech (protein/peptide) drugs, 71 nutraceuticals, and >3243 experimental drugs. Additionally, more than 2500 nonredundant protein (i.e., drug target) sequences are linked to these FDA

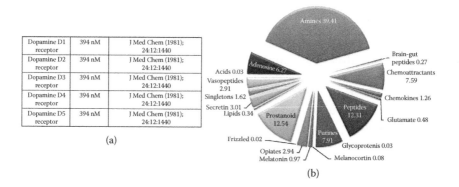

Dopamine D1 receptor	394 nM	J Med Chem (1981); 24:12:1440
Dopamine D2 receptor	394 nM	J Med Chem (1981); 24:12:1440
Dopamine D3 receptor	394 nM	J Med Chem (1981); 24:12:1440
Dopamine D4 receptor	394 nM	J Med Chem (1981); 24:12:1440
Dopamine D5 receptor	394 nM	J Med Chem (1981); 24:12:1440

(a)

(b)

Figure 6.2 Example of bias in biologically annotated drug-like compound data-bases. (a) ChEMBL binding data for compound CHEBI:137198 to dopamine receptors. The data are derived from a paper anterior to the identification of the five dopamine receptor subtypes and therefore reports the same affinity for the five subtypes. Hopefully, the data are assigned a confidence score of 0. The functional effect of the compound (inverse, partial-full agonist) is not documented and it could then be merged with a full antagonist in a QSAR model if only target name and binding data availability are used as filters. (b) Distribution (in percentage) of ligands from the MDL Drug Data Report (version 2007.1) for 19 out of 22 GPCR subfamilies. (From Weill, N. and Rognan, D., 2010, Alignment-Free Ultra-High-Throughput Comparison of Druggable Protein-Ligand Binding Sites, *J. Chem. Inf. Model.* 50:123–135.) Although the subfamily of biogenic amine receptors represents only 11% of all GPCRs, their ligands represent ca. 40% of known GPCR-binding molecules.

approved drug entries. Each DrugCard entry contains more than 100 data fields with half of the information devoted to drug/chemical data and the other half devoted to drug target or protein data. Two databases provide complementary information to DrugBank. The Therapeutic Target Database (TTD) (Zhu et al., 2010) permits similarity searches at the protein sequence level and precise functional effects of the ligands (agonist, antagonist, inhibitor). The SuperTarget database (Gunther et al., 2008) stores adverse drug effects.

6.2.3   Protein–ligand interaction databases

Knowledge bases describing target-ligand links can be divided into two categories depending on the nature of the association (structure of the complex, link in a network). The Protein Data Bank (PDB) (Berman et al., 2000) is the major source of structural information on protein–ligand complexes with ca. 68,000 entries out of which 48,000 describe a target-ligand complex. Since the PDB does not annotate ligands according to their biological properties, several dedicated databases focusing on proteins in complex with pharmacological ligands have emerged from the PDB. The screening-PDB database (sc-PDB) (Kellenberger et al., 2006) is the most exhaustive of them and stores 8198 entries comprising 2643 different proteins and 4236 different ligands. Druggability filters are used on both the ligand and binding site levels to select druggable protein–ligand complexes only. Structural databases provide very precise information but cover a limited biological space (2–3000 proteins) with respect to the druggable proteome. Complementary archives to the sc-PDB are the PDBbind (Wang et al., 2004) and BindingMOAD (Hu et al., 2005); databases which compile binding data (Kd, Ki, IC50) on protein–ligand complexes from the PDB (3214 for PDBbind and 4146 for BindingMOAD). Unfortunately, both databases have not been updated since 2008. Besides binding data repositories, interaction network databases represent a rapidly emerging field of research in chemogenomics and systems chemical biology in general. Drugs/chemicals are formally linked together or to targets, genes, pathways, diseases, and side effects according to known metabolic pathways, crystal structures, binding experiments, and drug–target relationships. STITCH (Kuhn et al., 2008) is the most exhaustive of this kind of database and registers interaction information for over 74,000 different chemicals (including 2200 drugs) and connects them to 2.5 million proteins in 630 organisms (see example in Figure 6.3).

Semanting Web technologies (Shadbolt and Berners-Lee, 2008) offer the possibility to integrate system chemical biology data from quite heterogeneous sources. Chem2Bio2RDF (Chen et al., 2010a) and SMO (Choi et al., 2010) have aggregated data for up to 25 different data sets classified into six categories (chemical and drug, protein and gene, chemogenomics, systems, phenotype, and literature), and enable queries for browsing high-dimensional spaces like polypharmacology, adverse drug reactions, or multiple pathway inhibition. Chemogenomic databases and tools are not only reserved to computational chemists for deriving statistical models or predicting protein–ligand binding. It is important to disseminate this information to any scientist working in the drug discovery area. Due to the complexity of the data, it is of utmost importance to release user-friendly graphical interfaces to browse chemogenomic space.

The graphical interface to STITCH (Kuhn et al., 2008) nicely permits jumping from the chemical to the target space and even to the interconnecting links (Figure 6.3). DrugViz (Xiong et al., 2008) is a plugin to Cytoscape, an open-source platform to display and mine drug–target networks. iPHACE (Garcia-Serna et al., 2010) is another visualizer recently described that permits display of annotated sdf file formats as interaction maps (*heatmaps*) and, for example, comparison of polypharmacological profiles (Figure 6.4).

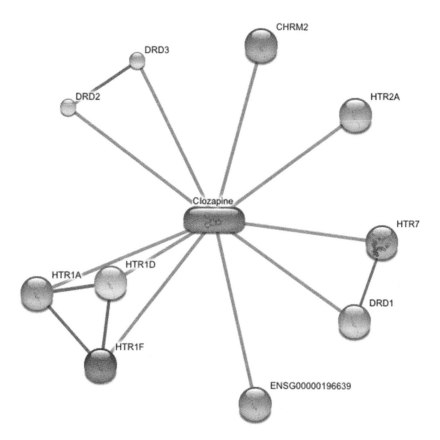

Figure 6.3 Interaction network around the antipsychotic drug clozapine. Drug–target interactions are shown in green and target–target interactions in blue. (Data taken from the STITCH2 Web site: http://stitch.embl.de/.) **(See color insert.)**

6.3   How to browse chemogenomic space?

There are three possible ways of querying and mining chemogenomic space depending on the nature of the object we are looking at. The analysis can be focused on bioactive ligands, bioactive targets, or target-ligand complexes. Depending on the point of view, different comparison algorithms and metrics will be used and reviewed from here on.

6.3.1   The ligand point of view

Two ligands with similar properties (structure and physicochemical descriptors) are more likely to share the same biological targets and biological properties (therapeutical indication, side effects) than two dissimilar ligands (Martin, Kofron, and Traphagen, 2002). Ligand

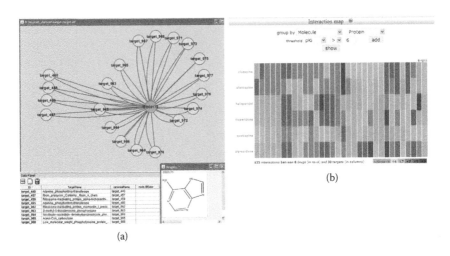

Figure 6.4 (a) Displaying the drug–target interaction network around adenine (DB00173) with DrugViz. (b) Interaction map of clozapine and five related compounds with 32 targets, generated with iPHACE. (From Garcia-Serna, R. et al., 2010, iPHACE: Integrative Navigation in Pharmacological Space, *Bioinformatics* 26:985–986.) **(See color insert.)**

similarity can therefore be considered an indirect measure of target similarity (Figure 6.5).

6.3.1.1 Predicting polypharmacology

One very important application of ligand-centric chemogenomic approaches is the prediction of all possible targets for a given ligand. Hence, many drugs bind to several macromolecular targets to achieve their therapeutical effect. For example, the antipsychotic drug clozapine is known to simultaneously bind to at least 16 different targets with sub-micromolar affinities (source: PDSP Ki database), a phenomenon called *polypharmacology* (Frantz, 2005). A recent analysis of ca. 5000 known interactions between 802 drugs and 480 targets reveals an average number of six different targets per drug (Mestres et al., 2009). *In silico*-driven target prediction is therefore important for selecting potential drug candidates with a well-defined polypharmacological profile, to anticipate side effects and adverse drug reactions of known drugs, or even for finding novel therapeutical indications to existing drugs (drug repurposing).

Three research groups have mainly contributed to establish the proof-of-concept that pairwise–ligand similarity searches are an efficient way of identifying targets for known ligands. The Novartis Institute for Biomolecular Research (NIBR) has pioneered the usage of ligand circular fingerprints to establish target-specific QSAR models by training machine learning algorithms (Naïve Bayes, Support Vector Machines) to

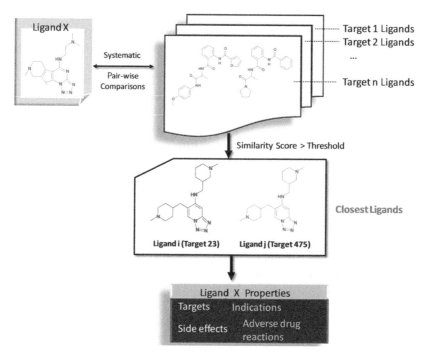

Figure 6.5 Ligand-centric approach to target and biological property annotation of a query molecule.

discriminate true actives from inactive compounds (Bender et al., 2006; Nettles et al., 2006; Nidhi et al., 2006). On average, the correct target was found 77% of the time when a multiple category Bayesian model was applied to external test sets (Nidhi et al., 2006).

Shoichet's group has developed the Similarity Ensemble Approach (SEA) (Keiser et al., 2007) in which a query ligand is compared to 246 target-specific ensembles of compounds (activity classes) by means of a simple Tanimoto coefficient measurement on Daylight topological fingerprints. The approach was successfully applied to predict the possible binding of 3665 FDA approved drugs to 246 MDL Drug Data Report (MDDR) targets (Keiser et al., 2009). Among all possible combinations, 30 were prioritized based on the statistical significance of the computed similarity scores. Twenty-three associations could be confirmed experimentally, out of which five were very potent (Ki < 100 nM) (Keiser et al., 2009). This remarkable achievement was however slightly biased by the relative low diversity of predicted targets (aminergic GPCRs) due to the known bias of biologically annotated MDDR ligands in aminergic GPCR ligands (recall Figure 6.2b). The SEA approach could later be applied to cross boundaries by predicting the possible binding of 746 commercial drugs to the protein farnesyltransferase (PFTase), a major target enzyme for developing antitumoral drugs (DeGraw et al., 2010). Among investigated drugs, the antiallergic drug Loratidine (H1 histamine antagonist) and the antifungal agent Miconazole (14α-demethylase inhibitor) were successfully predicted to bind to PFTase with low micromolar affinities (DeGraw et al., 2010).

Mestres et al. uses a conceptually similar but technically different approach based on Shannon Entropy Descriptors (SHED) computed from the distribution of atom-centered pharmacophoric feature pairs (Gregori-Puigjane and Mestres, 2006). This structural descriptor was used to annotate a library of 2000 nuclear hormone receptor (NHR) ligands covering 25 receptors, and to predict potential off-target NHR binding by a set of 3000 current drugs (Mestres et al., 2006). The approach was later extended to a larger set of ca. 700 therapeutically relevant targets (Gregori-Puigjane and Mestres, 2008b) and applied to predict the polypharmacological profile of 482 novel heterocyclic compounds against a restricted set of 86 GPCRs (Areias et al., 2010). Twenty-three of these target-orphan compounds received an annotation for at least one receptor, according to the corresponding pairwise similarity to known GPCR ligands. Since the four adenosine receptors (A_1, A_{2A}, A_{2B}, A_3) were frequent hitters, 10 ligands were screened *in vitro* for binding to the four adenosine receptors. Two of them were confirmed to be micromolar antagonists of the A_{2B} and A_3 receptors, respectively (Areias et al., 2010). Chemical similarity to both ligands led to 282 additional commercially available compounds, which were then virtually profiled against the four adenosine receptors. Of particular interest were eight compounds predicted to target the previously unreached A_1 and A_{2A} adenosine receptors. Three of them were conformed to be relatively selective micromolar antagonists of the later two receptor subtypes (Areias et al., 2010).

Due to the lack of exhaustive and public profiling data, the accuracy of ligand-based profiling methods is still unknown. A first answer was recently given by Vidal et al. (Vidal and Mestres, 2010), who tried to recover a recently published binding data matrix of 13 known antipsychotics on 34 targets (Roth et al., 2004). Relying on a reference set of 21,100 compounds with known binding affinities for the 34 targets, inhibition constants of the 13 drugs to the 34 targets were predicted from drug neighbors in the reference set close in chemical descriptor space by an inverse distance weighting interpolation. The scope of predictability as well as the recall was shown to rise with the binding affinity range whereas the precision was following the inverse relationships. Promisingly, 65% of all affinities could be predicted with a log unit error to a precision of 92% (Vidal and Mestres, 2010). This study highlights the very best of what *in silico* target profiling may reach for a set of drugs located in a very dense population of the drug-target chemogenomic space. However, the rapidly increasing availability of binding data on newer chemogenomic regions will undoubtedly enhance the applicability domain and precision of these *in silico* profiling methods.

The above-cited applications had been relying on carefully annotated and manually filtered ligand libraries (mostly DrugBank, MDR, and Wombat) (Table 6.1). Mining the larger but more noisy PubChem Bioassay data was thought to be more difficult but indeed proved to work reasonably well (although inferior to results obtained from handmade QSAR data) on a set of 655 bioassays when applying Naïve Bayesian modeling on circular fingerprints (Chen et al., 2010b). The accuracy of the models, as estimated by receiver operating characteristic (ROC) values, was much more dependent on the number of inactives than on the number of actives. For example, the accuracy of models based on primary HTS screens (with a lot of inactive compounds) was significantly higher than that derived from secondary follow-up assays.

6.3.1.2 Predicting adverse drug reactions

Since predicting potential targets by comparing target-annotated ligand sets is possible, the approach can be further coupled to the prediction of side effects and adverse drug reactions (ADRs) potentially linked to those targets. Hence, ADR was rated as the fourth leading cause of death in industrial countries (Giacomini et al., 2007). Preclinical safety profiling being quite expensive when realized at the experimental biochemical level, computational profiling is therefore gaining more and more importance in the pharmaceutical industry. The proof-of-concept was brought by Novartis in a series of three papers demonstrating that (i) ligand binding to a set of *hot* targets amenable to side effects can be predicted by multiple category Bayesian modeling with good confidence (Bender et al., 2007); (ii) ADR models can be drawn from the simple structure of known drugs with ADR annotation (Bender et al., 2007); and (iii) links between ADRs, their mediated targets and the implied metabolic pathways can be detected from properly annotated ligand sets (Scheiber et al., 2009b); ADR links could be mapped to common chemical features of causative drugs (Scheiber et al., 2009c).

6.3.1.3 Virtual screening of high-throughput screening (HTS) data

Ligand-centric chemogenomic screening can also be applied to detect noises in HTS data, notably cell-based assays, which are more prone to errors than biochemical assays. Novartis reported an analysis of 650,000 compounds tested on 13 gene reporter assays (Crisman et al., 2007). First, a multiple category Bayesian model was derived to predict molecular targets of frequent hitters (found active in three or more assays). Most of them were involved in apoptosis and cell differentiation (e.g., protein kinases) therefore proposing off-targets, whose recognition may perturb the bioassay. Interestingly, frequent hitters could not be detected by simple property distributions. The Bayesian frequent hitter model was last used to select 160 compounds from an external data set for experimental evaluation on a novel gene reporter assay. Fifty percent of predicted frequent hitters were recovered as active, demonstrating the accuracy of the initial model.

6.3.1.4   Library diversity analysis

On a prospective basis, ligand-centric target annotation of large compound libraries enables better definition of the current target coverage of existing ligands or drugs (Gregori-Puigjane and Mestres, 2008a) and detection of biases in compound libraries (Gregori-Puigjane and Mestres, 2008;a Mestres et al., 2008; Hert et al., 2009). Biases have been noted at the chemical level since screening libraries contain mostly compounds similar to existing metabolites and natural products (biogenic molecules) (Hert et al., 2009). On the other side, target space coverage was quite different according to the protein family. Whereas aminergic GPCRs are widely covered by existing ligands, only a small number of protein kinases concentrate the majority of compound annotations (Gregori-Puigjane and Mestres, 2008a). In-depth analysis of ligand similarity versus the sequence similarity of their corresponding targets indicate that ligand networks and target sequence networks differ significantly (Hert et al., 2008). In only 20% to 30% of the cases, sequence similarity corresponds to ligand similarity. Surprisingly, ligand networks are more natural and more organized according to network theory than their sequence network counterparts. Along the same line, sequence-based and ligand-based phylogenetic classifications of important target families (Vieth et al., 2004; Van der Horst et al., 2010) have been shown to be different, ligand-based classifications showing a higher level of disorder (from the protein point of view) simply reflecting a promiscuity not apparent from convergent target amino acid sequences.

6.3.2   The target point of view

Considering that similar targets bind similar ligands (Klabunde, 2007), a pure target-centric approach can be considered to identify ligands for a novel target (Figure 6.6). A simplistic approach is to compare amino acid sequences, which is usually enough to infer a biological function but not to predict ligand binding since binding site similarity does not necessarily follow sequence similarity. The most logical approach is to compare ligand cavities at a three-dimensional (3D) level, which inherently restricts the approach to the protein universe of known 3D structure. The Protein Data Bank (Berman et al., 2000), which stores publicly available 3D structures of macromolecules, currently stores over 68,000

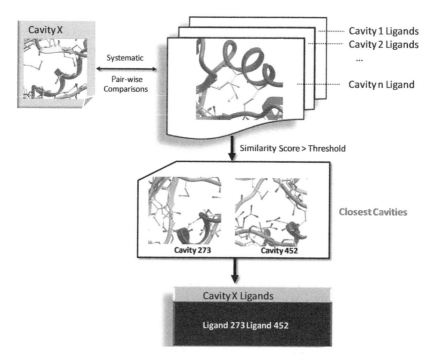

Figure 6.6 Target-centric approach to ligand annotation of a query target.

entries (September 2010 status). Further progress in comparative 3D modeling considerably extends the scope of protein structural space to 50,000 human protein models (Pieper et al., 2009). Last, outstanding efforts by structural genomic consortia to complete the structural proteome let us anticipate an acceptable coverage of the UniProt database (Schneider et al., 2009) in only 15 years (Nair et al., 2009). Cavity comparisons are therefore likely to play an important role in drug discovery, notably predicting structural druggability and ligand binding.

6.3.2.1 *Cavity detection and comparison methods*
The target-centric chemogenomic approach implies two prerequisites: (i) the ability to detect druggable ligand-binding cavities, and (ii) proper descriptors and metrics to assess ligand-binding site similarities.

The most simplistic approach is just to consider the sequence of binding site–lining residues at a protein family level (Surgand et al., 2006). Finding ligands for any given query–binding site is as simple as detecting the nearest cavities at the binding site sequence level among other family members and testing these ligands for binding to the query. This receptor hopping approach was followed to repurpose existing ligands for the CRTH2 (Frimurer et al., 2005) and GPCR6A (Faure et al., 2009) receptors, respectively.

Receptor hopping may even lead to subtype selective ligands after adequate medicinal chemistry optimization. Roche discovered a selective antagonist of the somatostatin type 5 receptor (SSTR5) by optimizing the binding properties of a histamine H1 antagonist lead (astemizole), SSTR5, and H1-binding cavities exhibiting similar physicochemical properties of a set on 35 consensus binding sites residues (Martin et al., 2007). This pragmatic approach is, however, limited to small target hops since a fixed number of binding site residues have to be compared. To find additional targets in a larger biological space, sequence and fold-independent comparison methods need to be used.

There are plenty of methods for automatically detecting cavities on protein surfaces (Perot et al., 2010). In most cases, druggable ligand-binding sites are among the top three biggest cavities at the protein surface (Huang, 2009) and are characterized by a well-defined range of topological and physicochemical properties (Halgren, 2009). A ligand-independent cavity detection approach presents the advantage of extending the applicability domain of cavity comparison tools to exosites (allosteric subsites, for example), but the drawback is to consider binding sites with no precedent ligand occupancy. Alternatively, a ligand-dependent strategy (Schalon et al., 2008) limits the scope of applicability but only to cavities with well-documented ligand occupancy. All described cavity comparison methods follow the same three-step flowchart. First, the structures of the two proteins to compare are parsed into meaningful 3D coordinates in order to reduce the complexity of the pairwise comparison. Typically, only key residues/atoms are considered and described by a limited number of points, which are labeled according to pharmacophoric, geometric, and/or chemical properties of their neighborhood. Second, the two resulting patterns are structurally aligned using, notably, clique detection (Kinoshita, Furui, and Nakamura, 2002; Schmitt, Kuhn, and Klebe, 2002) and geometric hashing methods (Shulman-Peleg, Nussinov, and Wolfson, 2004; Gold and Jackson, 2006) to identify the maximum number of equivalent points. Last, a scoring function quantifies the number of aligned features in the form of root-mean-square deviations (rmsd), residue conservation, or physicochemical property conservation. Pure shape-based methods have also been used to compute protein cavities and have led to exhaustive archives of protein–ligand binding sites (An, Totrov, and Abagyan, 2005; Weisel et al., 2009) that can be clustered or screened for envelope similarity to any query. Cavity comparison tools require a certain level of fuzziness to account for incomplete or erroneous side chain definitions and ligand-induced fit. In binding site descriptors, residue properties are therefore often mapped to Cα atoms coordinates (Yeturu and Chandra, 2008; Feldman and Labute, 2010; Weill and Rognan, 2010).

6.3.2.2 Binding site similarity searches to propose novel targets

One of the earliest, cavity-guided explanations of unexpected ligand cross-reactivity was described by Weber et al. (Weber et al., 2004). Starting from the observation that many cyclooxygenase type-2 (COX-2) inhibitors share with carbonic anhydrase (CA) inhibitors an arylsulfonamide moiety, known COX-2 inhibitors were tested for binding to various CA isoforms and revealed nanomolar binding affinities. A rational explanation to this cross-reactivity was obtained by comparing COX-2 inhibitor subpockets to a set of 9433 cavities from CavBase (Schmitt, Kuhn, and Klebe, 2002). For two of three subcavities, CA subpockets were retrieved among the top-scoring entries. However, no global similarity could be detected between entire ligand-binding pockets of both enzymes.

The Sequence Order-Independent Profile–Profile Alignment (SOIPPA) algorithm (Xie and Bourne, 2008) was designed and applied to detect similarity between unrelated proteins. Starting from a graph representation of protein Cα atoms with a geometric potential and position-specific amino acid score assigned to each graph node (Xie and Bourne, 2008), a maximum common subgraph is detected to align two proteins. The SOIPPA method was successfully used to predict remote binding site similarities between the binding site of selective estrogen receptor modulators (SERMs) at the Erα receptor and the Sarcoplasmic Reticulum Ca2 ion channel ATPase protein (SERCA) transmembrane domain (Xie, Wang, and Bourne, 2007), thus explaining known side effects of SERMs. Likewise, the NAD binding site of the Rossmann-fold and the S-adenosyl-methionine (SAM)-binding site of SAM-methyltransferases were found to be similar and consequently permitted to predict the cross-reactivity of catechol-O-methyltransferase inhibitors (entacapone, tolcapone) with the *M. tuberculosis* enoyl-acyl carrier protein reductase (Kinnings et al., 2009). The SOIPPA site comparison method was recently embedded in a flowchart involving subsequent docking to potential off-targets to remove potential false positives and integration of remaining off-targets in drug-target networks and metabolic pathways (Xie, Li, and Bourne, 2009; Chang et al., 2010).

Systematic pairwise comparison of the staurosporine-binding site of the proto-oncogene Pim-1 kinase with 6415 druggable protein–ligand binding sites (Kellenberger et al., 2006) using the SiteAlign algorithm (Schalon et al., 2008) suggested that the ATP-binding site of synapsin I (an ATP-binding protein regulating neurotransmitter release in the synapse) may recognize the pan-kinase inhibitor staurosporine (de Franchi et al., 2010). Biochemical validation of this hypothesis was realized by competition experiments of staurosporine with ATP-γ^{35}S for binding to synapsin I. Staurosporine, as well as three other inhibitors of protein kinases (cdk2, Pim-1, and casein kinase type 2), effectively bound to synapsin I with nanomolar affinities and promoted synapsin-induced F-actin bundling.

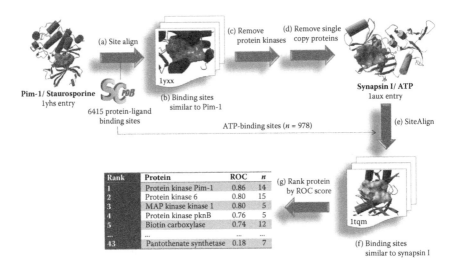

Figure 6.7 Computational protocol used to detect local similarities between ATP binding sites in Pim-1 kinase and synapsin I. (a) The ATP-binding site in Pim-1 kinase (occupied by the ligand staurosporine) is compared with SiteAlign (step a) to 6415 binding sites stored in the sc-PDB database. (From Schalon, C. et al., 2008, A Simple and Fuzzy Method to Align and Compare Druggable Ligand-Binding Sites, *Proteins* 71:1755–1778.) Among top scoring entries (step b), synapsin I is the only protein not belonging to the protein kinase target family (step c) and present in numerous copies (step d). A systematic SiteAlign comparison (e) of the ATP-binding site in synapsin I with 978 other ATP-binding sites suggest that some but not all ATP-binding sites of protein kinases (steps f and g) are similar to that of synapsin I. (From de Franchi, E. et al., 2010, Binding of Protein Kinase Inhibitors to Synapsin I Inferred from Pair-Wise Binding Site Similarity Measurements, *PLoS One* 5:e12214.)

The selective Pim-1 kinase inhibitor quercetagetin was shown to be the most potent synapsin I binder (IC_{50} = 0.15 μM), in agreement with the predicted binding site similarities between synapsin I and various protein kinases. Other protein kinase inhibitors (protein kinase A and chk1 inhibitor), kinase inhibitors (diacylglycerol kinase inhibitor), and various other ATP competitors (DNA topoisomerase II and HSP-90α inhibitors) did not bind to synapsin I, as predicted from a lower similarity of their respective ATP-binding sites to that of synapsin I (Figure 6.7).

Since some protein kinase inhibitors tested here actively and directly compete for the binding of ATP to synapsin I and modify the interactions of synapsin I with the actin-based cytoskeleton, it is tempting to speculate that at least some of the effects of protein kinase inhibitors on neurotransmitter release and presynaptic function are attributable to a direct binding to synapsin I and reveal new potential targets for the action of protein kinase inhibitors on synaptic transmission and plasticity.

Cavity comparisons can even lead to pocket deorphanization. The PocketPicker algorithm (Weisel, Proschak, and Schneider, 2007) was used to detect a cavity on the surface of an APOBEC3A structure, a protein able to inactivate retroviral genomes. Encoding the pocket as correlation vectors afforded comparison to a set of 1300 ligand-binding sites from the PDBbind data set (Wang et al., 2004). Among top scoring entries were only nucleic acid-binding pockets (Stauch et al., 2009). Point mutation of the cavity-lining residues effectively led to mutants with a reduced antiviral activity. The pocket was shown to recognize the small 5.8S RNA (Stauch et al., 2009) as a preliminary step to inactivate retroviral particles.

Predicting the polypharmacological profile of 17 inhibitors for protein kinases was also described by a cavity comparison approach (Milletti and Vulpetti, 2010). Using a fingerprint-based representation of ligand-binding pockets, the inhibitory profile to a set of 189 different protein kinases was predicted. Generally, kinase promiscuity was overpredicted but the target-based approach could retrieve as many true targets among the top scoring entries as a pure ligand-based approach relying on 2D or 3D descriptors of their corresponding inhibitors (Milletti and Vulpetti, 2010).

A conceptually similar but technically different solution to pocket similarity searches was recently illustrated by a pure pharmacophore-based strategy (Campagna-Slater and Schapira, 2009). Cavity residues of 10 different methyl-lysine binding domains (reading methyl-lysine signatures on histones) were first transformed in classical three-, four-, and five-feature pharmacophores. A collection of 22,568 PDB sites could be further screened by a standard pharmacophore search for entries verifying any of the query pharmacophores and yielded after various pruning steps a list of 236 binding sites. Five allosteric sites on known epigenetic targets could be selected for pocket similarity and suggest a potential role in epigenetic signaling.

The chemogenomic analysis is not restricted to comparing druggable-ligand binding sites. Davis and Sali (2010) reported an exhaustive alignment of overlapping protein–protein and protein–ligand binding sites to select cavities amenable to modulate protein–protein interaction with drug-like compounds. The analysis was able to recover known protein–protein interaction inhibitors and to propose novel interfaces that may be targeted by small molecular-weight compounds.

6.3.3 *The target-ligand point of view*

Navigating in chemogenomic space should logically be easier if both protein and ligand properties are considered simultaneously. Depending on whether 3D structures of protein–ligand complexes are available, three

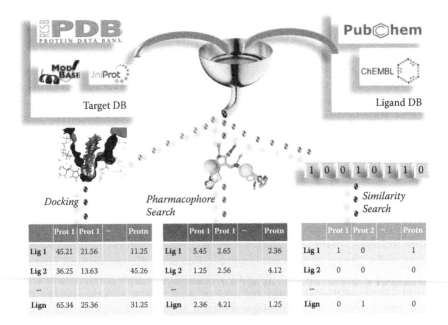

	Prot 1	Prot 1	...	Protn
Lig 1	45.21	21.56		11.25
Lig 2	36.25	13.63		45.26
...				
Lign	65.34	25.36		31.25

	Prot 1	Prot 1	...	Protn
Lig 1	5.45	2.65		2.36
Lig 2	1.25	2.56		4.12
...				
Lign	2.36	4.21		1.25

	Prot 1	Prot 2	...	Protn
Lig 1	1	0		1
Lig 2	0	0		0
...				
Lign	0	1		0

Figure 6.8 Integrated protein–ligand approaches to chemogenomic *in silico* screening. Ligands can be retrieved from public databases (e.g., PubChem, ChEMBL) and docked to protein 3D structures of interest (PDB: X-ray and NMR structures; ModBase: 3D comparative models) to yield protein–ligand docking matrices. Alternatively, protein–ligand complexes or models can be used as starting points to generate 3D pharmacophore databases and subsequently pharmacophore–ligand fitness matrices. Last, machine learning methods can be trained to discriminate true protein–ligand complexes (even of unknown 3D structures) using simple fingerprints embedding protein (or active site) and ligand structural information, and finally to yield binary protein–ligand association matrices.

different strategies with different strengths and limitations could be followed (Figure 6.8).

6.3.3.1 Parallel docking of ligand databases to protein databases

The current state-of-the art of reverse docking (docking a compound to a target collection) has recently been reviewed (Rognan, 2010). Since no major breakthroughs have happened in this field, we will just highlight the important underlying questions. First, parallel docking requires set-up of a collection of protein–ligand binding sites, a task which is far from being trivial when it has to be automated on a wide array of possible situations (cavities with or without ions, cofactors, prosthetic groups, covalently bound ligands, etc.). There are a few databases of protein–ligand binding sites that can be used for that purpose. The sc-PDB Database (http://bioinfo-pharma.u-strasbg.fr/scPDB) registers 8307 entries including 2682 unique proteins and 4289 unique ligands. It has benefited from several improvements over time (Kellenberger et al., 2006; Kellenberger, Foata, and Rognan, 2008) including enhanced biological annotation of protein targets, semiautomated correction of ligand atom types and bond orders, selection of druggable protein–ligand binding sites, definition of protein–ligand interaction fingerprints (IFPs) (Marcou and Rognan, 2007), and clustering of binding sites by physicochemical properties. Recently, the Potential Drug Target Database (http://www.dddc.ac.cn/pdtd/) was developed along the same ideas from PDB targets of known drugs; it currently includes 1027 entries for 847 unique targets (Gao et al., 2008). Last, the BioDrugScreen resource storing over 3,000,000 protein–ligand docking solutions (poses, scores) originates from the docking of the NCI diversity set (1600 compounds) to a panel of 1589 human targets (Li, Li et al., 2010). The docking matrix was later exploited to find compounds sharing the virtual profile of erlotinib, a known EGFR kinase inhibitor. Twelve compounds were identified, among which some also inhibited EGFR and displayed strong antitumor properties in lung tumor cells (Li, Bum-Erdene et al., 2010). For an exhaustive report of successful docking-based target identification, the reader is invited to a recent review (Rognan, 2010).

A docking-based approach presents the advantage of proposing a protein–ligand recognition model and thus guiding hit to lead optimization. However, docking is notoriously a slow computational procedure (ca. 30 s/compound) and amenable to serious errors in estimating binding free energies (Ferrara et al., 2004). Although topological postdocking approaches relying on interaction fingerprints (Deng, Chuaqui, and Singh, 2004; Marcou and Rognan, 2007; Kellenberger, Foata, and Rognan, 2008; Das, Krein, and Breneman, 2010; Sato, Honma, and Yokoyama, 2010) have been proposed to correct scoring function deficiencies, docking is clearly not the method of choice for massively profiling ligand collections.

6.3.3.2 *Pharmacophore-based parallel screening*

Instead of docking compounds to protein–ligand binding sites, it is faster to just check whether the ligand fulfills pharmacophore features derived from protein–ligand complexes. This concept was brought first by Langer et al. (Steindl et al., 2006) for profiling bioactive ligands against a small collection of structure-based pharmacophores. Starting from protein–ligand complexes of known X-ray structure, pharmacophore hypotheses are generated by an automated pharmacophore perception algorithm (Wolber and Langer, 2005) and converted into pharmacophore queries. The protocol was validated on a set of known protein–ligand complexes (Steindl et al., 2006) spanning various activity classes, embedded into an automated workflow and successfully used to assign three targets to two secondary metabolites from the medicinal plant *Ruta graveolens* (Rollinger et al., 2009). Currently, more than 2500 pharmacophore models are available for parallel virtual screening as part of the Inte:Ligand pharmacophore database. It covers 300 unique clinically relevant pharmacological targets originating from major therapeutical classes as well as antitargets such as the human *Ether-à-go-go* Related Gene (hERG), and members of the cytochrome P450 family.

The PharmMapper server (Liu et al., 2010) offers a free resource to fish targets by a reverse pharmacophore search. It encompasses a database of 7000 protein–ligand-derived pharmacophores (PharmTargetDB) and a fast ligand-to-pharmacophore fitting tool enabling a full screening within 1 to 2 h. The method was only validated on the classical examples of tamoxifen and methotrexate to permit retrieval of true targets of those two drugs among the top scorers. Complex-based pharmacophore models may be complemented by classical ligand-based pharmacophores to augment the scope of applicability, notably to protein classes (e.g., membrane proteins) for which few experimental structures but numerous ligands are available.

6.3.3.3 *Protein–ligand fingerprint similarity searches*

Contrarily to reverse docking and pharmacophore searches, protein–ligand fingerprints can be encoded in very simple bitstrings or real vectors, and do not necessitate true 3D information.

6.3.3.3.1 Fingerprinting interacting substructures and fragments In a series of papers, Bajorath et al. reported a way to augment classical ligand fingerprints with protein–ligand interaction-derived information (Tan, Batista, Bajorath, 2010a). The basic underlying idea is that all atoms/ groups of a bioactive ligand are not equally responsible for their biological activity. Focusing a chemical fingerprint to protein-interacting ligand atoms is likely to enhance the value of such a fingerprint by avoiding the possibility of recruiting novel compounds by a pure ligand-based virtual screening approach for the wrong reasons (atoms/groups not contributing to the cavity).

In the Interaction Annotated Structural Features (IASF) method, a typical ligand 2D fingerprint is annotated with energy-based weights derived from X-ray structures (Crisman, Sisay, and Bajorath, 2008), thus giving more importance to protein-interacting ligand substructures (called *interacting fragment* or *IF*). Such IF-annotated fingerprints were shown to outperform conventional fingerprints in standard similarity searches to known ligands of diverse activity classes. The IF itself can be encoded as a key-type fingerprint and used as a reference for similarity searches, often with superior performance to full ligand-based fingerprints (Tan, Lounkine, and Bajorath, 2008). Due to the very different complexity levels of full ligands and IF fingerprints, the advantage of using a simplified

IF substructure representation vanishes if low-complexity databases are screened (Tan, Batista, Bajorath, 2010b). IF information could be transferred from target A inhibitors to fingerprinting inhibitors of a related target B and improve recovery of true target B inhibitors by similarity search (Tan and Bajorath, 2009). Fingerprint scaling according to the frequency of occurrence of IFs, in other words, how frequent a substructural motif interacts with a target protein, provided another advantage to classical fingerprints in conventional similarity searches (Tan et al., 2009). To avoid the generation of artificial IF keys not present in whole molecules, thus biasing results, IFs were encoded at the atomic level (Batista, Tan, and Bajorath, 2010). Such fingerprints are, however, only applicable to ligands with a conserved binding mode to their targets. In case of different or overlapping binding modes (e.g., HIV-1 reverse transcriptase inhibitors), consensus interacting fragments cannot be simply derived and the IF fingerprinting method should thus be avoided (Tan, Batista, Bajorath, 2010b). With the increasing availability of protein–ligand 3D structures, protein–ligand interaction databases like CREDO (Schreyer and Blundell, 2009) or SMID (Snyder et al., 2006) will provide a good starting point to integrate protein–ligand interactions in a novel generation of fingerprints for chemogenomic screening.

6.3.3.3.2   Fingerprinting ligands and target descriptors   Three studies on fingerprinting GPCR–ligand pairs in chemogenomic applications have been described (Bock and Gough, 2005; Jacob et al., 2008; Weill and Rognan, 2009). In a pioneering work, Bock et al. (Bock and Gough, 2005) used rather standard 2D topological and atomic descriptors for ligands, physicochemical properties of amino acid sequences for receptors, and concatenate feature vectors for both the receptor and the ligand in a single fingerprint. A support vector machine (SVM) model was trained on 5319 receptor–ligand pairs from the PDSP Ki database (Roth, Sheffler, and Kroeze, 2004) to predict the Ki of any ligand to any GPCR and used to propose novel ligands for orphan GPCRs. Unfortunately, none of these predictions have been validated up to now. In the fingerprint proposed by Weill et al. (Weill et al., 2009), ligand properties have been represented by standard descriptors (MACCS keys and SHED descriptors), while protein cavities are encoded by a fixed-length bitstring describing pharmacophoric properties of a fixed number of binding site residues. Several machine learning classification algorithms (SVM, Random Forest, Naive Bayes) were trained on two sets of roughly 200,000 receptor–ligand fingerprints with a different definition of inactive decoys. Cross-validated models show excellent precision (>0.9) in distinguishing true from false pairs with a particular preference for Random Forest models. In most cases, predicting ligands for a given receptor was easier than predicting receptors for a given ligand.

Protein and ligand descriptors need not be concatenated into a single fingerprint. Different kernels for ligands and for targets may be developed as proposed by Vert et al. (Jacob and Vert, 2008; Jacob et al., 2008; Vert and Jacob, 2008), the similarity in target-ligand space being just the tensor product of ligand and target similarities, in support vector machine (SVM) classification models. Interestingly, the most precise protein kernel (based on binding site similarities, for example,) is not necessarily the most reliable. Bajorath and colleagues (Wassermann, Geppert, and Bajorath, 2009) suggested that simplified strategies for designing target-ligand SVM kernels should be used since varying the complexity of the target kernel does not influence much the identification of ligands for virtually deorphanized targets. Hence, predicting protein–ligand association is clearly dominated by ligand neighborhood (Wassermann, Geppert, and Bajorath, 2009).

Strömbergsson et al. proposed a generic amino acid sequence-based local descriptor for the proteome (Strömbergsson et al., 2008). Each residue of the protein is encoded by a set of nonoverlapping residue fragments describing its neighborhood. Ligands are classically described by a set of 27 diverse Dragon descriptors (Virtual Computational Chemistry Laboratory: http://www.vcclab.org/lab/edragon/). Starting from a set of 1421 triplets (protein, ligand, pKi inhibition constant) of known X-ray structure, the inhibition constant of 542 test complexes of unknown structures was predicted by SVM regression with a promising accuracy ($r^2 = 0.53$, root-mean-square error of prediction of 1.50) with regard to the diversity of the test set. The proteochemometric approach was extended to the analysis of the structural protein–ligand PDB space and of the druggable DrugBank-target space (Strömbergsson and Kleywegt, 2009). Both chemogenomic spaces were shown to present a minimal level of overlap since the PDB is biased with enzyme inhibitors whereas DrugBank is biased with existing drugs (mainly membrane receptor ligands). A better merge of both spaces in a single data set was recently reported (Strömbergsson et al., 2010) in which 7000 protein–ligand inhibition constants were gathered from the BindingDB database (Liu et al., 2007) classified either as high (pKi > 7) or low (pki < 7). Targets were described by 167 sequence-based PROFEAT descriptors (Li et al., 2006) and ligands by Dragon descriptors. Diverse machine learning algorithms (SVM, decision trees, Naïve Bayes) were trained on 50% of the data, validated on 25% remaining, and finally tested for the last 25% data used as an external data set. The best model was obtained from a decision tree approach and led to an acceptable accuracy of 81.6% for the validation set and of 75% on the external set. In contradiction to previous studies (Wassermann, Geppert, and Bajorath, 2009; Weill and Rognan, 2009), a pure ligand-based model led to accuracies and ROC curves typical from random selection, raising questions about the suitability of chosen ligand descriptors. Hence, the utility of embedding target information into a chemogenomic classification vanishes rapidly when enough ligand information is already available for describing a target. In agreement with a previous work (Jacob and Vert, 2008) on three target classes (GPCRs, enzymes, ion channels), our own experience with 7000 druggable protein–ligand complexes from the Protein Data Bank suggests using chemogenomic models only if less than 30–50 ligands are available for each target (Meslamani and Rognan, 2010). Moreover, 3D binding site descriptors did not outperform simple sequence-based descriptors when a chemogenomic model was necessary (Meslamani and Rognan, 2010).

Although these results seem disappointing (2D ligand similarity searches also usually outperform 3D similarity searches), it enables the application of chemogenomic methods to a very large set of targets of unknown 3D structures. What still has to be done is to select a target-dependent strategy (ligand-based, chemogenomic-based) to optimize the screening efficiency. The continuous growth of target-annotated ligand databases should boost this tendency in the very near future.

6.4   Conclusions

The current explosion in publicly available chemogenomic databases has initiated novel computational strategies to predict protein–ligand interaction matrices. A main conflict associated with this novel computational chemistry method is that most data are biased to some overrepresented activity classes (e.g., biogenic amine receptor antagonists, ion channel blockers) and targets, for which precise 3D structural information is missing. Conversely, targets with the highest level of 3D structural information (enzymes from the Protein Data Bank) are represented with fewer copies of not really drug-like ligands (often transition state inhibitors) as a mirror of poorer history in drug discovery programs. Despite these limitations, computational chemogenomics is a field of intense research showing, along with the increasing availability of public data, significant progress and achievements, which would have been inconceivable a decade ago. We foresee within the next 10 years their application to a very large majority of druggable human proteins and a revolution in computational chemist minds similar to what medicinal chemists are currently experiencing with the advent of fragment-based drug discovery.

References

An, J., Totrov, M., and Abagyan, R. 2005. Pocketome via comprehensive identification and classification of ligand binding envelopes. *Mol. Cell. Proteomics* 4:752–761.

Apweiler, R., Bairoch, A., Wu, C. H., et al. 2004. UniProt: The Universal Protein knowledgebase. *Nucleic Acids Res.* 32: D115–119.

Areias, F. M., Brea, J., Gregori-Puigjane, E., et al. 2010. *In silico* directed chemical probing of the adenosine receptor family. *Bioorg. Med. Chem.* 18: 3043–3052.

Bajorath, J. 2008. Computational analysis of ligand relationships within target families. *Curr. Opin. Chem. Biol.* 12:352–358.

Batista, J., Tan, L., and Bajorath, J. 2010. Atom-centered interacting fragments and similarity search applications. *J. Chem. Inf. Model.* 50:79–86.

Bender, A., Jenkins, J. L., Glick, M., et al. 2006. "Bayes affinity fingerprints" improve retrieval rates in virtual screening and define orthogonal bioactivity space: When are multitarget drugs a feasible concept? *J. Chem. Inf. Model.* 46:2445–2456.

Bender, A., Scheiber, J., Glick, M., et al. 2007. Analysis of pharmacology data and the prediction of adverse drug reactions and off-target effects from chemical structure. *ChemMedChem* 2:861–873.

Berman, H. M., Westbrook, J., Feng, Z., et al. 2000. The Protein Data Bank. *Nucleic Acids Res.* 28:235–242.

Bock, J. R. and Gough, D. A. 2005. Virtual screen for ligands of orphan G protein-coupled receptors. *J. Chem. Inf. Model.* 45:1402–1414.

Campagna-Slater, V. and Schapira, M. 2009. Evaluation of virtual screening as a tool for chemical genetic applications. *J. Chem. Inf. Model.* 49:2082–2091.

Campbell, S. J., Gaulton, A., Marshall, J., et al. 2010. Visualizing the drug target landscape. *Drug Discov. Today* 15:3–15.

Caron, P. R., Mullican, M. D., Mashal, R. D., et al. 2001. Chemogenomic approaches to drug discovery. *Curr. Opin. Chem. Biol.* 5: 464–470.

Chang, R. L., Xie, L., Bourne, P. E., and Palsson, B. O. 2010. Drug off-target effects predicted using structural analysis in the context of a metabolic network model. *PLoS Comput. Biol.* 6: e1000938.

Chen, B., Dong, X., Jiao, D., et al. 2010a. Chem2Bio2RDF: A semantic framework for linking and data mining chemogenomic and systems chemical biology data. *BMC Bioinformatics* 11:255.

Chen, B. and Wild, D. J. 2010b. PubChem bioassays as a data source for predictive models. *J. Mol. Graph. Model.* 28:420–426.

Chen, B., Wild, D., and Guha, R. 2009. PubChem as a source of polypharmacology. *J. Chem. Inf. Model.* 49: 2044–2055.

Choi, J., Davis, M. J., Newman, A. F., and Ragan, M. A. 2010. A semantic web ontology for small molecules and their biological targets. *J. Chem. Inf. Model.* 50:732–741.

Crisman, T. J., Parker, C. N., Jenkins, J. L., et al. 2007. Understanding false positives in reporter gene assays: *In silico* chemogenomics approaches to prioritize cell-based HTS data. *J. Chem. Inf. Model.* 47:1319–1327.

Crisman, T. J., Sisay, M. T., and Bajorath, J. 2008. Ligand-target interaction-based weighting of substructures for virtual screening. *J. Chem. Inf. Model.* 48:1955–1964.

Daniel, M., Stuart, L., Christopher, C., Stuart, W., and Adam, N. 2009. Synthesis of natural-product-like molecules with over eighty distinct scaffolds. *Angew. Chem. Int. Ed. Engl.* 48:104–109.

Das, S., Krein, M. P., and Breneman, C. M. 2010. Binding affinity prediction with property-encoded shape distribution signatures. *J. Chem. Inf. Model.* 50:298–308.

Davis, F. P. and Sali, A. 2010. The overlap of small molecule and protein binding sites within families of protein structures. *PLoS Comput. Biol.* 6:e1000668.

de Franchi, E., Schalon, C., Messa, M., et al. 2010. Binding of protein kinase inhibitors to synapsin I inferred from pair-wise binding site similarity measurements. *PLoS One* 5:e12214.

DeGraw, A. J., Keiser, M. J., Ochocki, J. D., Shoichet, B. K., and Distefano, M. D. 2010. Prediction and evaluation of protein farnesyltransferase inhibition by commercial drugs. *J. Med. Chem.* 53:2464–2471.

Deng, Z., Chuaqui, C., and Singh, J. 2004. Structural interaction fingerprint (SIFt): A novel method for analyzing three-dimensional protein-ligand binding interactions. *J. Med. Chem.* 47:337–344.

Dessailly, B. H., Nair, R., Jaroszewski, L., et al. 2009. PSI-2: Structural genomics to cover protein domain family space. *Structure* 17:869–881.

Faure, H., Gorojankina, T., Rice, N., et al. 2009. Molecular determinants of noncompetitive antagonist binding to the mouse GPRC6A receptor. *Cell. Calcium* 46:323–332.

Feldman, H. J. and Labute, P. 2010. Pocket similarity: Are alpha carbons enough? *J. Chem. Inf. Model.* 50:1466–1475.

Ferrara, P., Gohlke, H., Price, D. J., Klebe, G., and Brooks, C. L., 3rd. 2004. Assessing scoring functions for protein-ligand interactions. *J. Med. Chem.* 47:3032–3047.

Frantz, S. 2005. Drug discovery: Playing dirty. *Nature* 437:942–943.

Frimurer, T. M., Ulven, T., Elling, C. E., et al. 2005. A physicogenetic method to assign ligand-binding relationships between 7TM receptors. *Bioorg. Med. Chem. Lett.* 15:3707–3712.

Gao, Z., Li, H., Zhang, H., et al. 2008. PDTD: A Web-accessible protein database for drug target identification. *BMC Bioinformatics* 9:104.

Garcia-Serna, R., Ursu, O., Oprea, T. I., and Mestres, J. 2010. iPHACE: Integrative navigation in pharmacological space. *Bioinformatics* 26:985–986.

Giacomini, K. M., Krauss, R. M., Roden, D. M., et al. 2007. When good drugs go bad. *Nature* 446:975–977.

Gold, N. D. and Jackson, R. M. 2006. Fold independent structural comparisons of protein-ligand binding sites for exploring functional relationships. *J. Mol. Biol.* 355:1112–1124.

Gregori-Puigjane, E. and Mestres, J. 2006. SHED: Shannon entropy descriptors from topological feature distributions. *J. Chem. Inf. Model.* 46:1615–1622.

Gregori-Puigjane, E. and Mestres, J. 2008a. Coverage and bias in chemical library design. *Curr. Opin. Chem. Biol.* 12:359–365.

———. 2008b. A ligand-based approach to mining the chemogenomic space of drugs. *Comb. Chem. High. Throughput Screen.* 11:669–676.

Gunther, S., Kuhn, M., Dunkel, M., et al. 2008. SuperTarget and matador: Resources for exploring drug-target relationships. *Nucleic Acids Res.* 36:D919–922.

Halgren, T. A. 2009. Identifying and characterizing binding sites and assessing druggability. *J. Chem. Inf. Model.* 49:377–389.

Harmar, A. J., Hills, R. A., Rosser, E. M., et al. 2009. IUPHAR-DB: The IUPHAR database of G protein-coupled receptors and ion channels. *Nucleic Acids Res.* 37:D680–685.

Hert, J., Irwin, J. J., Laggner, C., Keiser, M. J., and Shoichet, B. K. 2009. Quantifying biogenic bias in screening libraries. *Nat. Chem. Biol.* 5:479–483.

Hert, J., Keiser, M. J., Irwin, J. J., Oprea, T. I., and Shoichet, B. K. 2008. Quantifying the relationships among drug classes. *J. Chem. Inf. Model.* 48:755–765.

Hu, L., Benson, M. L., Smith, R. D., Lerner, M. G., and Carlson, H. A. 2005. Binding MOAD (mother of all databases). *Proteins* 60:333–340.

Huang, B. 2009. MetaPocket: A meta approach to improve protein ligand binding site prediction. *OMICS* 13:325–330.

Hur, J. and Wild, D. J. 2008. PubChemSR: A search and retrieval tool for PubChem. *Chem. Cent. J.* 2:11.

Jacob, L., Hoffmann, B., Stoven, V., and Vert, J. P. 2008. Virtual screening of GPCRs: An *in silico* chemogenomics approach. *BMC Bioinformatics* 9:363.

Jacob, L. and Vert, J. P. 2008. Protein-ligand interaction prediction: An improved chemogenomics approach. *Bioinformatics* 24:2149–2156.

Keiser, M. J., Roth, B. L., Armbruster, B. N., et al. 2007. Relating protein pharmacology by ligand chemistry. *Nat. Biotechnol.* 25:197–206.

Keiser, M. J., Setola, V., Irwin, J. J., et al. 2009. Predicting new molecular targets for known drugs. *Nature* 462:175–181.

Kellenberger, E., Foata, N., and Rognan, D. 2008. Ranking targets in structure-based virtual screening of three-dimensional protein libraries: Methods and problems. *J. Chem. Inf. Model.* 48:1014–1025.

Kellenberger, E., Muller, P., Schalon, C., et al. 2006. sc-PDB: An annotated database of druggable binding sites from the Protein Data Bank. *J. Chem. Inf. Model.* 46:717–727.

Kinnings, S. L., Liu, N., Buchmeier, N., et al. 2009. Drug discovery using chemical systems biology: Repositioning the safe medicine Comtan to treat multi-drug and extensively drug resistant tuberculosis. *PLoS Comput. Biol.* 5:e1000423.

Kinoshita, K., Furui, J., and Nakamura, H. 2002. Identification of protein functions from a molecular surface database, eF-site. *J. Struct. Funct. Genomics* 2:9–22.

Klabunde, T. 2007. Chemogenomic approaches to drug discovery: Similar receptors bind similar ligands. *Br. J. Pharmacol.* 152:5–7.

Klabunde, T. and Evers, A. 2005. GPCR antitarget modeling: Pharmacophore models for biogenic amine binding GPCRs to avoid GPCR-mediated side effects. *Chembiochem.* 6:876–889.

Kuhn, M., von Mering, C., Campillos, M., Jensen, L. J., and Bork, P. 2008. STITCH: Interaction networks of chemicals and proteins. *Nucleic Acids Res.* 36:D684–688.

Li, L., Bum-Erdene, K., Baenziger, P. H., et al. 2010. BioDrugScreen: A computational drug design resource for ranking molecules docked to the human proteome. *Nucleic Acids Res.* 38:D765–773.

Li, L., Li, J., Khanna, M., et al. 2010. Docking small molecules to predicted off-targets of the cancer drug erlotinib leads to inhibitors of lung cancer cell proliferation with suitable *in vitro* pharmacokinetic properties. *ACS Med. Chem. Lett.* 1:229–233.

Li, Q., Wang, Y., and Bryant, S. H. 2009. A novel method for mining highly imbalanced high-throughput screening data in PubChem. *Bioinformatics* 25:3310–3316.

Li, Z. R., Lin, H. H., Han, L. Y., et al. 2006. PROFEAT: A web server for computing structural and physicochemical features of proteins and peptides from amino acid sequence. *Nucleic Acids Res.* 34:W32–37.

Liu, T., Lin, Y., Wen, X., Jorissen, R. N., and Gilson, M. K. 2007. BindingDB: A web-accessible database of experimentally determined protein-ligand binding affinities. *Nucleic Acids Res.* 35:D198–201.

Liu, X., Ouyang, S., Yu, B., et al. 2010. PharmMapper server: A web server for potential drug target identification using pharmacophore mapping approach. *Nucleic Acids Res.* 38 Suppl: W609–614.

Marcou, G. and Rognan, D. 2007. Optimizing fragment and scaffold docking by use of molecular interaction fingerprints. *J. Chem. Inf. Model.* 47:195–207.

Martin, R. E., Green, L. G., Guba, W., Kratochwil, N., and Christ, A. 2007. Discovery of the first nonpeptidic, small-molecule, highly selective somatostatin receptor subtype 5 antagonists: A chemogenomics approach. *J. Med. Chem.* 50:6291–6294.

Martin, Y. C., Kofron, J. L., and Traphagen, L. M. 2002. Do structurally similar molecules have similar biological activity? *J. Med. Chem.* 45:4350–4358.

Meslamani, J. E. and Rognan, D. 2010. Predicting Ligand-Protein Binding Using SVM: A Novel Chemogenomic Screening Method. 18th EUROQSAR workshop, Rhodes, Greece, 2010:117.

Mestres, J., Gregori-Puigjane, E., Valverde, S., and Sole, R. V. 2008. Data completeness—The Achilles heel of drug-target networks. *Nat. Biotechnol.* 26:983–984.

———. 2009. The topology of drug-target interaction networks: Implicit dependence on drug properties and target families. *Mol. Biosyst.* 5:1051–1057.

Mestres, J., Martin-Couce, L., Gregori-Puigjane, E., Cases, M., and Boyer, S. 2006. Ligand-based approach to *in silico* pharmacology: Nuclear receptor profiling. *J. Chem. Inf. Model.* 46:2725–2736.

Milletti, F. and Vulpetti, A. 2010. Predicting polypharmacology by binding site similarity: From kinases to the protein universe. *J. Chem. Inf. Model.* 50:1418–1431.

Nair, R., Liu, J., Soong, T. T., et al. 2009. Structural genomics is the largest contributor of novel structural leverage. *J. Struct. Funct. Genomics* 10:181–191.

Nettles, J. H., Jenkins, J. L., Bender, A., et al. 2006. Bridging chemical and biological space: "Target fishing" using 2D and 3D molecular descriptors. *J. Med. Chem.* 49:6802–6810.

Nidhi, Glick, M., Davies, J. W., and Jenkins, J. L. 2006. Prediction of biological targets for compounds using multiple-category Bayesian models trained on chemogenomics databases. *J. Chem. Inf. Model.* 46:1124–1133.

Nielsen, T. E. and Schreiber, S. L. 2008. Towards the optimal screening collection: A synthesis strategy. *Angew. Chem. Int. Ed. Engl.* 47:48–56.

Ong, S. E., Schenone, M., Margolin, A. A., et al. 2009. Identifying the proteins to which small-molecule probes and drugs bind in cells. *Proc. Natl. Acad. Sci. USA* 106:4617–4622.

Overington, J. 2009. ChEMBL. An Interview with John Overington, Team Leader, Chemogenomics at the European Bioinformatics Institute Outstation of the European Molecular Biology Laboratory (EMBL-EBI). Interview by Wendy A. Warr. *J. Comput. Aided Mol. Des.* 23:195–198.

Paolini, G. V., Shapland, R. H., Van Hoorn, W. P., Mason, J. S., and Hopkins, A. L. 2006. Global mapping of pharmacological space. *Nat. Biotechnol.* 24:805–815.

Perot, S., Sperandio, O., Miteva, M. A., Camproux, A. C., and Villoutreix, B. O. 2010. Druggable pockets and binding site centric chemical space: A paradigm shift in drug discovery. *Drug Discov. Today* 15:656–667.

Pieper, U., Eswar, N., Webb, B. M., et al. 2009. MODBASE, a database of annotated comparative protein structure models and associated resources. *Nucleic Acids Res.* 37:D347–354.

Rognan, D. 2007. Chemogenomic approaches to rational drug design. *Br. J. Pharmacol.* 152:38–52.

Rognan, D. 2010. Structure-based approaches to target fishing and ligand profiling. *Mol. Inf.* 29:176–187.

Rollinger, J. M., Schuster, D., Danzl, B., et al. 2009. *In silico* target fishing for rationalized ligand discovery exemplified on constituents of *Ruta graveolens*.
Planta Med. 75:195–204.

Roth, B. L., Sheffler, D. J., and Kroeze, W. K. 2004. Magic shotguns versus magic
bullets: selectively nonselective drugs for mood disorders and schizophrenia. *Nat. Rev. Drug Discov.* 3:353–359.

Sato, T., Honma, T., and Yokoyama, S. 2010. Combining machine learning and
pharmacophore-based interaction fingerprint for *in silico* screening. *J. Chem.
Inf. Model.* 50:170–185.

Schalon, C., Surgand, J. S., Kellenberger, E., and Rognan, D. 2008. A simple and
fuzzy method to align and compare druggable ligand-binding sites. *Proteins*
71:1755–1778.

Scheiber, J., Chen, B., Milik, M., et al. 2009a. Gaining insight into off-target mediated effects of drug candidates with a comprehensive systems chemical biology analysis. *J. Chem. Inf. Model.* 49:308–317.

Scheiber, J. and Jenkins, J. L. 2009b. Chemogenomic analysis of safety profiling
data. *Methods Mol. Biol.* 575:207–223.

Scheiber, J., Jenkins, J. L., Sukuru, S. C., et al. 2009c. Mapping adverse drug reactions in chemical space. *J. Med. Chem.* 52:3103–3107.

Schierz, A. C. 2009. Virtual screening of bioassay data. *J. Cheminform.* 1:21.

Schmidtke, P. and Barril, X. 2010. Understanding and predicting druggability. A
high-throughput method for detection of drug binding sites. *J. Med. Chem.*
53:5858–5867.

Schmitt, S., Kuhn, D., and Klebe, G. 2002. A new method to detect related function
among proteins independent of sequence and fold homology. *J. Mol. Biol.*
323:387–406.

Schneider, M., Lane, L., Boutet, E., et al. 2009. The UniProtKB/Swiss-Prot knowledgebase and its plant proteome annotation program. *J. Proteomics* 72:567–573.

Schreyer, A. and Blundell, T. 2009. CREDO: A protein-ligand interaction database
for drug discovery. *Chem. Biol. Drug Des.* 73:157–167.

Schuster, D., Laggner, C., and Langer, T. 2005. Why drugs fail—A study on side
effects in new chemical entities. *Curr. Pharm. Des.* 11:3545–3559.

Shadbolt, N. and Berners-Lee, T. 2008. Web science emerges. *Sci. Am.* 299:76–81.

Shoichet, B. K. 2004. Virtual screening of chemical libraries. *Nature* 432:862–865.

Shoshan, M. C. and Linder, S. 2008. Target specificity and off-target effects as determinants of cancer drug efficacy. *Expert. Opin. Drug Metab. Toxicol.* 4:273–280.

Shulman-Peleg, A., Nussinov, R., and Wolfson, H. J. 2004. Recognition of functional sites in protein structures. *J. Mol. Biol.* 339:607–633.

Snyder, K. A., Feldman, H. J., Dumontier, M., Salama, J. J., and Hogue, C. W. 2006.
Domain-based small molecule binding site annotation. *BMC Bioinformatics* 7:152.

Southan, C., Varkonyi, P., and Muresan, S. 2009. Quantitative assesment of the
expanding complementarity between public and commercial databases of
bioactive compounds. *J. Cheminform.* 1:10.

Stauch, B., Hofmann, H., Perkovic, M., et al. 2009. Model structure of APOBEC3C
reveals a binding pocket modulating ribonucleic acid interaction required
for encapsidation. *Proc. Natl. Acad. Sci. USA* 106:12079–12084.

Steindl, T. M., Schuster, D., Laggner, C., and Langer, T. 2006. Parallel screening: A
novel concept in pharmacophore modeling and virtual screening. *J. Chem.
Inf. Model.* 46:2146–2157.

Strömbergsson, H., Daniluk, P., Kryshtafovych, A., et al. 2008. Interaction model based on local protein substructures generalizes to the entire structural enzyme-ligand space. *J. Chem. Inf. Model.* 48:2278–2288.

Strömbergsson, H. and Kleywegt, G. J. 2009. A chemogenomics view on protein-ligand spaces. *BMC Bioinformatics* 10 Suppl 6:S13.

Strömbergsson, H., Lapins, M., Kleywegt, G. J., and Wikberg, J. E. S. 2010. Towards proteome-wide interaction models using the proteochemometrics approach. *Mol. Inf.* 29:499–508.

Surgand, J. S., Rodrigo, J., Kellenberger, E., and Rognan, D. 2006. A chemogenomic analysis of the transmembrane binding cavity of human G-protein–coupled receptors. *Proteins* 62:509–538.

Tan, L. and Bajorath, J. 2009. Utilizing target-ligand interaction information in fingerprint searching for ligands of related targets. *Chem. Biol. Drug Des.* 74:25–32.

Tan, L., Batista, J., and Bajorath, J. 2010a. Computational methodologies for compound database searching that utilize experimental protein-ligand interaction information. *Chem. Biol. Drug Des.* 76:191–200.

———. 2010b. Rationalization of the performance and target dependence of similarity searching incorporating protein-ligand interaction information. *Che. Inf. Model.* 50:1042–1052.

Tan, L., Lounkine, E., and Bajorath, J. 2008. Similarity searching using fingerprints of molecular fragments involved in protein–ligand interactions. *J. Chem. Inf. Model.* 48:2308–2312.

Tan, L., Vogt, M., and Bajorath, J. 2009. Three-dimensional protein-ligand interaction scaling of two-dimensional fingerprints. *Chem. Biol. Drug Des.* 74:449–456.

Ursu, O. and Oprea, T. I. 2010. Model-free drug-likeness from fragments. *J. Chem. Inf. Model.* 50:1387–1394.

Van der Horst, E., Peironcely, J. E., Ijzerman, A. P., et al. 2010. A novel chemogenomics analysis of G protein-coupled receptors (GPCRs) and their ligands: A potential strategy for receptor de-orphanization. *BMC Bioinformatics* 11:316.

Vert, J. P. and Jacob, L. 2008. Machine learning for *in silico* virtual screening and chemical genomics: New strategies. *Comb. Chem. High Throughput Screen.* 11:677–685.

Vidal, D. and Mestres, J. 2010. *In silico* receptorome screening of antipsychotic drugs. *Mol. Inf.* 29:543–551.

Vieth, M., Higgs, R. E., Robertson, D. H., et al. 2004. Kinomics-structural biology and chemogenomics of kinase inhibitors and targets. *Biochim. Biophys. Acta* 1697:243–257.

Wang, R., Fang, X., Lu, Y., and Wang, S. 2004. The PDBbind database: Collection of binding affinities for protein-ligand complexes with known three-dimensional structures. *J. Med. Chem.* 47:2977–2980.

Wang, Y., Xiao, J., Suzek, T. O., et al. 2009. PubChem: A public information system for analyzing bioactivities of small molecules. *Nucleic Acids Res.* 37:W623–633.

Wassermann, A. M., Geppert, H., and Bajorath, J. 2009. Ligand prediction for orphan targets using support vector machines and various target-ligand kernels is dominated by nearest neighbor effects. *J. Chem. Inf. Model.* 49:2155–2167.

Weber, A., Casini, A., Heine, A., et al. 2004. Unexpected nanomolar inhibition of carbonic anhydrase by COX-2-selective celecoxib: New pharmacological opportunities due to related binding site recognition. *J. Med. Chem.* 47:550–557.

Weill, N. and Rognan, D. 2009. Development and validation of a novel protein-ligand fingerprint to mine chemogenomic space: Application to G protein-coupled receptors and their ligands. *J. Chem. Inf. Model.* 49:1049–1062.

Weill, N. and Rognan, D. 2010. Alignment-free ultra-high-throughput comparison of druggable protein-ligand binding sites. *J. Chem. Inf. Model.* 50:123–135.

Weisel, M., Proschak, E., Kriegl, J. M., and Schneider, G. 2009. Form follows function: Shape analysis of protein cavities for receptor-based drug design. *Proteomics* 9:451–459.

Weisel, M., Proschak, E., and Schneider, G. 2007. PocketPicker: Analysis of ligand binding-sites with shape descriptors. *Chem. Cent. J.* 1:7.

Wermuth, C. G. 2006. Similarity in drugs: Reflections on analogue design. *Drug Discov. Today* 11:348–354.

Wishart, D. S., Knox, C., Guo, A. C., et al. 2008. DrugBank: A knowledgebase for drugs, drug actions and drug targets. *Nucleic Acids Res.* 36:D901–906.

Wolber, G. and Langer, T. 2005. Ligand Scout: 3-D pharmacophores derived from protein-bound ligands and their use as virtual screening filters. *J. Chem. Inf. Model.* 45:160–169.

Xie, L. and Bourne, P. E. 2007. A robust and efficient algorithm for the shape description of protein structures and its application in predicting ligand binding sites. *BMC Bioinformatics* 8 Suppl 4:S9.

———. 2008. Detecting evolutionary relationships across existing fold space, using sequence order-independent profile-profile alignments. *Proc. Natl. Acad. Sci. USA* 105:5441–5446.

Xie, L., Li, J., and Bourne, P. E. 2009. Drug discovery using chemical systems biology: Identification of the protein-ligand binding network to explain the side effects of CETP inhibitors. *PLoS Comput. Biol.* 5: e1000387.

Xie, L., Wang, J., and Bourne, P. E. 2007. *In silico* elucidation of the molecular mechanism defining the adverse effect of selective estrogen receptor modulators. *PLoS Comput. Biol.* 3:e217.

Xiong, B., Liu, K., Wu, J., et al. 2008. DrugViz: A Cytoscape plugin for visualizing and analyzing small molecule drugs in biological networks. *Bioinformatics* 24:2117–2118.

Yang, L., Chen, J., and He, L. 2009. Harvesting candidate genes responsible for serious adverse drug reactions from a chemical-protein interactome. *PLoS Comput. Biol.* 5:e1000441.

Yeturu, K. and Chandra, N. 2008. PocketMatch: A new algorithm to compare binding sites in protein structures. *BMC Bioinformatics* 9:543.

Zhu, F., Han, B., Kumar, P., et al. 2010. Update of TTD: Therapeutic Target Database. *Nucleic Acids Res.* 38:D787–791.

Glossary

Aptamer Aptamers are oligonucleic acid or peptide molecules that bind to a specific target molecule. Aptamers are usually created by selecting them from a large random sequence pool, but natural aptamers also exist in riboswitches. Aptamers can be combined with ribozymes to self-cleave in the presence of their target molecule. DNA or RNA aptamers consist of (usually short) strands of oligonucleotides, while peptide aptamers consist of a short variable peptide domain, attached at both ends to a protein scaffold.

Bioassay A bioassay is a procedure for determining the concentration, purity, and/or biological activity of a substance (e.g., vitamin, hormone, plant growth factor, antibiotic, enzyme) by measuring its effect on an organism, tissue, cell, enzyme, or receptor preparation compared to a standard preparation.

Bioinformatics The first level can be defined as the design and application of methods for the collection, organization, indexing, storage, and analysis of biological sequences (both nucleic acids and proteins). The next stage of bioinformatics is the derivation of knowledge concerning the pathways, functions, and interactions of these genes and proteins.

Biomarker(s) A molecular indicator of a specific biological property: a biochemical feature or facet that can be used to measure the progress of disease or the effects of treatment. Biomarker substances sometimes found in an increased amount in the blood, other body fluids, or tissues indicate the presence of the disease.

Biology-Oriented Synthesis Biology-oriented synthesis (BIOS) is a synthetic design approach, which aims to expand the chemical space around scaffolds of natural origin.

Biosensor A device that uses specific biochemical reactions mediated by isolated enzymes, immune system, tissues, organelles, or whole cells to detect chemical compounds, usually by electrical, thermal, or optical signals.

Cellular Assays Live cell assays fully automate sample handling and quantify cellular characteristics such as motility, proliferation, and morphology. This functional assay technology is amenable to high-throughput analysis, and therefore it is complementary to many proteomic technologies focused on identification of potential therapeutic targets.

Chemical Biology Chemical biology utilizes small molecules as a tool to resolve biological problems or to understand the roles of proteins by interfering or by regulating the biological processes. Chemical biology is closely related to chemical genetics at the genomics level. Chemical biology offers the potential to enhance the drug discovery process and as a result of integration of biology and chemistry in characterizing biochemical pathways associated with disease states.

Chemical Diversity The *unrelatedness* of a set of, for example, building blocks or members of a combinatorial library, as measured by their properties such as atom connectivity, physical properties, computational measurements, or bioactivity. Diverse chemical libraries are widely and evenly dispersed in the multidimensional property space. Chemical diversity has two major elements, redundancy and coverage.

Chemical Genetics Chemical genetics describes the use of small synthetic molecules that elicit a phenotypic change by direct protein interaction, to identify key genes involved in a specific biological pathway of interest. In many cases existing drugs are used as the chemical probes whose overall effect is well established but whose mode of action is not well understood.

Chemical Genomics The current wealth of gene sequence information available, in conjunction with recent advances in array-based technologies, have facilitated a chemical genomic approach. High-throughput screening with small, highly specific synthetic molecules, against multiple protein targets, results in measurable phenotypic changes and presents an opportunity to do gene functional analysis and create new therapeutic leads at the same time.

Chemical Microarrays A diverse set of small molecules immobilized on a chip surface is a highly efficient tool for the study of precious biological samples.

Chemical Proteomics The general concept of annotating function to new proteins by discovering small molecule ligands might be referred to as chemical proteomics. To link new proteins with known catalytic activities, proteome-scale screens for generic enzyme activities (e.g., protease and phosphatase) are implemented as major tools.

Chemogenomics A bioinformatics-based alternative to screening, based on a target family design of specific inhibitors, ligands. *Chemogenomics* is often used to describe the focused exploration of target gene families, in which small molecule leads—identified by virtue of their interaction with a single member of a gene family—are used to study the biological role of other members of that family, the function of which is unknown.

Chemoproteomics Drug discovery technology that evaluates the global interactions of small drug-like molecules with all proteins in a proteome or subproteome as a means to identify new protein targets and to elucidate protein function, biochemical pathways, and protein reaction partners.

Clustering Organizing a set of compounds into clusters is one standard approach to assess the diversity of libraries. Clusters are a group of compounds, which are related by structural or behavioral properties.

Combinatorial Chemistry A high-throughput chemical approach using a combinatorial process to prepare sets of compounds from sets of different building blocks that form a combinatorial library.

Combinatorial Library A set of compounds prepared by combinatorial chemistry where the building blocks are connected to each other in all possible variations, frequently around a molecular skeleton or scaffold.

Combinatorial Synthesis A process to prepare large sets of organic compounds by combining sets of building blocks.

Diversity-Oriented Synthesis Diversity-oriented synthesis (DOS) is a strategy for quick access to molecule libraries with an emphasis on skeletal diversity. In DOS strategy divergent synthesis aims to generate a library of chemical compounds by first reacting a multifunctional molecule with a set of various types of reactants leading to different skeletons or chemotypes.

Docking Computational techniques for the exploration of the possible binding modes of a substrate to a given receptor, enzyme, or other binding site.

Druggability Druggability is a term used in drug discovery to describe the suitability of a protein or protein complex to be targeted by a drug or drug-like molecule, in a way that this interaction will alter the protein's function and correct disease-causing behavior.

Epigenetics Epigenetics is the study of heritable changes in gene expression or cellular phenotype caused by mechanisms other than changes in the underlying DNA sequence—hence the name *epi-* (Greek επί- meaning over, above, outer) *genetics.* It refers to functionally relevant modifications to the genome that do not involve a change in the nucleotide sequence. Examples of such changes are DNA methylation and histone deacetylation.

Forward Chemical Genetics Forward chemical genetics involves identifying a phenotype in an organism or cell caused by a small molecule and then identifying the target affected. In principle, this is analogous to a classical genetics screen, in which one screens for a mutation that has a desired phenotype and then identifies the mutant gene that is responsible.

Functional Genomics The development and application of experimental approaches to assess gene function by making use of the information and reagents provided by structural genomics.

Gene Expression Profiling A technology used to identify the differential expression patterns of known genes in model systems, including disease states.

Genomics The study of gene expression, structure, and function of all genes of a specific organism.

Genotype The genetic makeup of an individual organism is the genetic makeup of a cell, an organism, or an individual (i.e., the specific allele makeup of the individual) usually with reference to a specific character under consideration.

Genotyping A process of determining differences in the genetic makeup (genotype) of an individual by examining the individual's DNA sequence using biological assays and comparing it to another individual's sequence or a reference sequence. It reveals the alleles an individual has inherited from their parents. Traditionally, genotyping is the use of DNA sequences to define biological populations by use of molecular tools. It does not usually involve defining the genes of an individual.

Glycomics Glycomics is the comprehensive study of glycomes (the entire complement of sugars, whether free or present in more complex molecules, of an organism), including genetic, physiologic, pathologic, and other aspects. Glycomics "is the systematic study of all glycan structures of a given cell type or organism" and is a subset of glycobiology. (*Essentials of Glycobiology*, 2nd ed., 2009, Cold Spring Harbor Laboratory Press.)

G-Protein Coupled Receptor The members of the superfamily of G-protein coupled receptors (GPCRs) consist of seven transmembrane regions with different extra- and intracellular domains. The GPCRs mediate signals through a membrane, when activated binding a G-protein.

High-Throughput Screening Automated methods to test chemical compounds for *in vitro* efficacy. HTS is capable of screening large numbers of compounds prior to being tested in animals. It can flag novel compounds, but provides limited information as to their effect on living animals.

Homology Modeling A computational process that uses one or more 3D structures of homologous proteins to predict the unknown 3D structure of a sequentially related protein.

Human Genome Project Human Genome Project is the global initiative to map and sequence the human genome. The completion of the draft human genome sequence was published as the draft human genome sequence in *Nature,* 15 February 2001.

Immobilization The immobilization chemistry allows for covalently or noncovalently attaching small molecules or biopolymers onto a glass microscope slide. This method produces uniform spots containing equivalent amounts of samples.

Immunomics Immunomics is the study of the molecular functions associated with all immune-related coding and noncoding mRNA transcripts. To unravel the function, regulation and diversity of the immunome requires that we identify and correctly categorize all immune-related transcripts. The importance of intercalated genes, antisense transcripts, and noncoding RNAs and their potential role in regulation of immune development and function are only just starting to be appreciated.

In Silico Literally *in the computer.* It can be used to screen out compounds which are not druggable. Narrower terms include *in silico* biology, *in silico* modeling, *in silico* proteomics, *in silico* screening, and so forth.

In Silico **Methods** Computational analogs of the methods carried out *in vitro* and *in vivo.* For example, metabolic stability can be tested *in vitro* and predicted by *in silico* methods.

Interactome Interactome is defined as the whole set of molecular interactions in cells. Databases describing molecular interactions and genetic networks will provide a link at the functional level between the genome, the proteome, and the transcriptome worlds of different organisms. Interaction databases therefore aim at describing the contents, structure, function, and behavior of what we herein define as the interactome world.

Lead Discovery The process of identifying active new chemical entities, which by subsequent modification may be transformed into a clinically useful drug.

Lead Generation Strategies developed to identify compounds which possess a desired but nonoptimized biological activity.

Lead Optimization The synthetic modification of a biologically active compound, to fulfill all stereoelectronic, physicochemical, pharmacokinetic, and toxicologic requirements for clinical usefulness.

Lipidome The lipidome is defined as the total lipid composition of a cell or organ. The lipidome includes but is not limited to fatty acids, acylglycerols, glycerophospholipids, sphingolipids, terpenes, steroids, and lipids that complex with proteins and polysaccharides.

Lipidomics Lipidomics means a systematic profiling of lipids and the entities that interact with lipids. Lipidomics may be extended to the genomics of lipid metabolism, understanding the biosynthesis pathways, lipoproteins, and proteins that interact and metabolize lipids.

Medicinal Chemistry A chemistry-based discipline, also involving aspects of biological, medical, and pharmaceutical sciences. It is concerned with the invention, discovery, design, identification, and preparation of biologically active compounds, the study of their metabolism, the interpretation of their mode of action at the molecular level, and the construction of structure–activity relationships.

Metabolome A metabolome is the quantitative complement of all the low-molecular weight molecules present in cells in a particular physiological or developmental state.

Metabolomics A comprehensive and quantitative analysis of all metabolites that helps researchers understand biological systems, since the effect of a single mutation often leads to the alteration of the metabolite level of seemingly unrelated biochemical pathways. Since such an analysis reveals the changes in the metabolome of the biological system (composition of the metabolites), this approach is called *metabolomics*.

Microarray A microscopic, ordered array of nucleic acids, proteins, small molecules, cells, or other substances that enables parallel analysis of complex biochemical samples. The term *microarray* originally referred to spotted cDNA arrays, but now others and we use it for any hybridization-based array. The major characteristics of microarrays are that they have a glass or plastic slide as a matrix, and are created with robots that deposit probes on the slides.

Microfluidics Lab-on-a-chip technology based on the transport of nanoliter or picoliter volumes of fluids through microchannels within a glass or plastic chip. Microfluidics systems evolved from MEMS research.

Omics The *-ome* suffix refers to cellular and molecular biology, forming nouns with the sense *all constituents considered collectively*. Omics studies the corresponding *ome* as a whole (*en masse*).

Pharmacogenomics Pharmacogenomics is broadly defined as the study of interindividual variations in whole-genome or candidate gene single-nucleotide polymorphism maps, haplotype markers, and alterations in gene expression that might correlate with drug responses.

Phenotype Phenotype is an observable characteristic displayed by an organism. These characteristics can be controlled by genes, by the environment, or by a combination of both. The characteristic can be directly observable, such as having brown eyes. In some cases, the phenotype will be measurable, such as having high blood pressure.

Phenotypic Screen The systematic classification and characterization of phenotypes are essential for ultimately mapping the genes responsible for normal and abnormal development and physiology. In any search for mutations or altered functional expression, identification depends on phenotypic screening and its ability to detect variation from normal.

Polypharmacology Polypharmacology, which studies the interaction of a drug with multiple targets, provides a new way to address the issue of high attrition rates arising from lack of efficacy and toxicity. The investigation of polypharmacology can be done through *in silico* or *in vivo* activity mapping leading to a biological network.

Protein Families Sets of proteins that share a common evolutionary origin reflected by their relatedness in function, which is usually reflected by similarities in sequence, or in primary, secondary, or tertiary structure. Subsets of proteins with related structure and function.

Proteomics Proteomics is the study of protein expression, structure, and function, and the interactions of all proteins of a specific organism.

Reverse Chemical Genetics A way to discover a useful chemical tool/probe is to start with the desired protein target and screen for small molecules that affect its activity, then ask whether the small molecule causes a phenotypic change in an organism or cell. This approach is analogous to reverse genetics, in which a gene is deliberately mutated or knocked out in order to study the resulting phenotype. Reverse chemical genetics has been extensively used in the pharmaceutical industry and leads to better biological understanding.

Structure–Activity Relationship Structure–activity relationship (SAR) is the analysis of the quantitative relationship between the biological activity of a set of compounds and their spatial properties using statistical methods.

Scaffold Scaffold is the core portion of a molecule common to all members of a combinatorial library.

Screening Pharmacological or toxicological screening consists of a specified set of procedures to which a series of compounds is subjected to characterize pharmacological and toxicological properties and to establish dose–effect and dose–response relationships.

Sequence Alignment A linear comparison of amino (or nucleic) acid sequences, in which insertions are made in order to bring equivalent positions in adjacent sequences into the correct register. Alignments are the basis of sequence analysis methods, and are used to pinpoint the occurrence of conserved motifs.

Signaling Pathways These are biochemical pathways by which cells are communicating with each other and regulating cell behavior and function. They operate through specific protein–protein interactions or utilize second messenger small molecules.

Signal Transduction Signal transduction occurs when an extracellular signaling molecule activates a cell surface receptor. In turn, this receptor alters intracellular molecules, creating a response. There are two stages in this process: a signaling molecule activates a specific receptor on the cell membrane and then a second messenger transmits the signal into the cell, leading to a physiological response.

Single Nucleotide Polymorphism When comparing the same sequence from two individuals, there can often be single base pair changes or single nucleotide polymorphisms (SNPs). These can be useful genetic markers.

Solid Phase Synthesis Solid phase synthesis is synthesis of compounds on the solid surface of an insoluble support such as tiny plastic beads.

Solution Phase Parallel Synthesis Strategy whereby sets of discrete compounds are prepared simultaneously in solution in arrays of physically separate reaction vessels or microcompartments without interchange of intermediates during the assembly process. Solution phase techniques utilize several thousand organic synthetic methods directly.

Steroidomics Steroidomics is profiling the free sterol content of cerebrospinal fluid by a combination of charge tagging and liquid chromatography–tandem mass spectrometry.

Structure-Based Design Structure-based design is a design strategy for new chemical entities based on the three-dimensional (3D) structure of the target obtained by X-ray or nuclear magnetic resonance (NMR) studies, or from protein homology models.

Systems Biology Systems biology is a term used to describe a number of trends in bioscience research, and a movement that draws on those trends. Proponents describe systems biology as a biology-based interdisciplinary study field that focuses on complex interactions in biological systems.

Target A molecule (usually a protein gene product, but sometimes a DNA sequence, or, in the case of antisense drugs, an mRNA) that may interact with a drug or drug candidate.

Target Identification Identifying a molecule (often a protein) that is instrumental to a disease process (though not necessarily directly involved), with the intention of finding a way to regulate that molecule's activity for therapeutic purposes.

Target Validation The association of a gene and/or its gene product as a molecular target for drug intervention in a particular pathology.

Toxicogenomics Toxicogenomics is a field of science that deals with the collection, interpretation, and storage of information about gene and protein activity within particular cells or tissues of an organism in response to toxic substances. Toxicogenomics combines toxicology with genomics or other high-throughput molecular profiling technologies such as transcriptomics, proteomics, and metabolomics.

Transcription Factor A protein that binds to regulatory regions and helps control gene expression.

Virtual Screening Selection of compounds by evaluating their desirability in a computational model.

X-Ray Crystallography Atomic resolution experimental structure determination method that uses the diffraction pattern of X-rays on crystals to determine the structure.

Index

Printed and bound by CPI Group (UK) Ltd, Croydon, CR0 4YY

21/10/2024

01777105-0001

—